Decision Making in Uncertain Situations:
An Extension to the Mathematical Theory of Evidence

by

Fabio Campos

ISBN: 1-58112- 335-3

DISSERTATION.COM

Boca Raton, Florida
USA • 2006

Decision Making in Uncertain Situations:
An Extension to the Mathematical Theory of Evidence

Dissertation.com
Boca Raton, Florida
USA • 2006

ISBN: 1-58112- 335-3

Decision Making in Uncertain Situations

An Extension to the Mathematical Theory of Evidence

Fábio Campos

Recife

December 2005

The danger of dreaming is to achieve the dreams.

This work is a fruit of an old dream.

I dedicate it:

to everyone who chases their dreams,

to Márcia, my dream,

and to my father Ryam (in memoriam) who, from my earliest years taught me this and the appreciation for research and experimentation in all fields of life.

iv

Acknowledgements

Every sacrifice or effort to do what one likes becomes, in the end, a great delight.

Thus, if there is someone who deserves the most thanks for the innumerable nights and weekends without me, this person would be my wife, Márcia, who gave up the most for this work, followed closely by the rest of my family, in particular, my mom, Hermínia, my grandmother, Dulce, and my friends.

I also could not forget to thank:

— my friends from the Informatics Center - UFPE, in particular those from the GrECo (Computacional Engineering Group), companions of many night-long journeys of studies and debates to solve the most diverse computer problems;

— to my graduate tutor, Prof. Sérgio Cavalcante, who always thinks of himself as a problem solver, overcoming all the obstacles to the development of the ideas presented here, believing and supporting this cool madness that allowed ourselves create a work in an area totally diverse from which that which we initially conceived, with all the risks associated;

— to André Leite (Graduate student in the Master's program) who patiently helped with the revising of many details;

— and to Prof. Fernando Campello de Souza an unconditional supporter of all my scientific initiatives, both academic and entrepreneurial, to whom I owe a significant portion of my formation, and more precisely that part which has helped me to create the present work.

What it is the reality around us but the perception that we extract from the probabilities?

What it is the "known universe" but a huge amount of empty space, populated here and there by regions of probable encounters among subatomic particles whose behavior we can perceive as solids, liquids and gasses?

And indeed what are people, things, ideas and thoughts but probabilistically-coded information, being constantly updated for the next event, implying in things like their progress, movement, elaboration, decrease and growth?

That is, everything, in fact, consists of an eternal play of the divine dice...

— FÁBIO CAMPOS (2004)

Contents

List of Tables

Foreword

The core of this work was presented to the Computer Science Graduation Program of the Informatics Center at "Universidade Federal de Pernambuco – UFPE" as one of the requirements for the obtention of a PhD in Computer Science.

Before we pass to the content of the book itself it is important to explain the motivation behind it. The work from which this book is a consequence began in the Computation Engineering and Embedded Systems area and finished as a PhD thesis in the area of Uncertain Probabilities [7].

Our initial project goal was to achieve "A Methodology for Embedded Systems Design", where we intended to study knowledge representation derived from the user inputs, the representation of uncertainty and conflict in a knowledge-base, and a mapping algorithm that would have allowed us to associate users' requisites to the project orientations suggested by the knowledge-base.

To implement this project, there was the need to adopt a knowledge representation model able to consider both the uncertainty and the conflict present in the inputs supplied by the users and those existing internally in the knowledge-base. Among the several available formalisms for the representation and combination of this kind of knowledge, we chose, for technical matters, mainly the kind of sample space and the nature of the uncertainties involved, Evidence Theory, also known as the Dempster-Shafer Theory, or the Mathematical Theory of Evidence.

When we began to study the Dempster-Shafer Theory, we verified that theoretical problems could cause counter-intuitive results or limit the modeling power of this formalism, making the practical application of this theory, in our particular application, excessively complicated. In the search of a better understanding of the problem, we verified that these issues had been already suggested by one of the creators of the theory, Glenn Shafer, in his seminal article presenting the theory itself, dated 1976 [45]. In this article, Shafer explain that counter-intuitive results would occur when the evidence to be combined had belief concentration in disjoints events and a only small quantity of belief in a common event. Although there have been several attempts to solve this behavior, only partially acceptable solutions have been achieved. Through a simple but ingenious solution, we managed to solve this classical problem, also obtaining as a by-product a more epistemic modeling for combining all kinds of evidence (that is, not just in cases where the subject is known to show counter-intuitive results before hand). This extension of the Dempster-Shafer Theory allows its use in a larger range of applications, as well as providing an improvement in the modeling of the original range of applications.

Thus, the solution of this classical problem, together with its implications, became the object of my PhD thesis. When we introduced this solution to the Decision Theory Group of UFPE, the questions and discussions raised helped us see that the conceptual implications of the solution had ramifications rather larger than those we had glimpsed initially; and this led us to what became the main object of this book, an extension to the Mathematical Theory of Evidence, able to treat both objective and epistemic uncertainties, allowing decision making in uncertain situations [8].

Organization of the Book

The initial chapter introduces the main goal of this book and provides the establishment of a framework that allows us to treat either the subjective as the objective uncertainties, solving the classical problems of the theories that deal with evidence, as well as eliminating the dichotomy of treatment between these two kinds of uncertainties. This is followed by the justification and motivation which provides the foundation for a work of this kind.

The following chapter describes the state-of-the-art and related works. The first section of this chapter is devoted to the subject of subjective uncertainty showing its philosophical basis and the need to consider this kind of uncertainty. It continues with an explanation of knowledge representation and combination, in Section 2.3, where the several kinds of imperfections to which information can be subjected are shown. Section 2.4 follows with an explanation of several interpretations of "probability". In this section, several different nuances of interpretation are clarified, which also provides the basis for the title of the book: "Decision Making in Uncertain Situations – An Extension to the Mathematical Theory of Evidence", since it deals with beliefs and not with "classic probabilities".

Chapter 3 explains the Dempster-Shafer Theory, the conceptual and formal basis upon which the extension presented in this book is built.

Chapter 4 proposes an including beliefs theory. Section 4.1 clarifies imperfections of the Dempster-Shafer Theory, and Section 4.2 proposes a solution, through the adoption of a new rule of evidence combination and the associated conceptual framework (vide Section 4.2.2). This section concludes with a validation of the extension to the theory (Section 4.2.3).

Chapter 5 presents a case study of a possible application of the extension. This example elucidates the importance of the quantification of the uncertainty embodied in the results using real financial data from the Brazilian market.

Concluding the book, Chapter 6 summarizes the results and conclusions obtained, as well their practical and conceptual implications.

1 Introduction

This work extends the Mathematical Theory of Evidence, also known as the Dempster-Shafer Theory, through the adoption of a new rule for the combination of evidence and a companion set of concepts. This extension solves the counter-intuitive problems illustrated in the original theory, extends its power of expression and allows the representation of uncertainty in the results.

The representation of uncertainty implies the possibility of its use in decision-making and also makes explicit the relationship between the numeric results achieved and the results from classical probability theory.

The main problem addressed by this book is how to model and combine bodies of knowledge while maintaining the representation of the unkowledge and of the conflict among the bodies.

This is a problem with far-reaching applications in many knowledge segments, in particular for the field of artificial intelligence. It must be kept in mind that knowledge based systems depend on algorithms able to relate the inputs of a system to a correct answer coming out of the knowledge-base, and both the inputs and the knowledge-base are subject to information imperfections caused by the unknowledge and the conflict.

There are several formalism to deal with knowledge representation and combination, among them the Mathematical Theory of Evidence or Dempster-Shafer Theory. This theory has received considerable attention since it is suitable to represent naturally a wide range of situations [42] and is more general than the widely-used Bayesian Theory, which relates to a specific case of the Theory of Evidence [25].

However, the Mathematical Theory of Evidence exhibits two major limitations: the possibility of arriving at a counter-intuitive result; and the lack of representation of the degree of subjective uncertainty in the results [21]. This latter limitation can permit the combination of conflicting bodies of knowledge resulting in the same numeric values as a combination of bodies with no conflict at all. These limitations narrow the range of applications of the theory while at the same time demand the elimination of some bodies of knowledge which could, in other ways, contribute to the construction of the knowledge. It should be noted that the original theory presents as a principle that if a source of knowledge is consulted, the evidence coming from it must be taken into consideration, even if only to increase the degree of uncertainty in the combination result (the "specialist" concept) [45].

Since the formalization of the theory by Shafer in 1976 [45], several attempts have been made to identify the cause and solution of counter-intuitive behavior resulting from several particular combinations. For example, Reference [67] represents a collection of papers in which such authors as Smets, Yager and Zadeh attribute counter-intuitive behavior to the normalization step of Dempster's Rule of Combination. Based on this premise several other rules of combination have been tried, which eliminate the normalization step (vide Section 4.2.3 for references and authors) in an attempt to solve at least part of the counter-intuitive behavior of the original rule. The adoption of these rules, however, leads to important side effects that also result in counter-intuitive behaviors in some specific situations.

Taking all of this into account, the main objective of this book is to present an extension to the Theory of Dempster-Shafer by the development of a conceptual basis and the adoption of a new rule of evidence combination able to increase significantly the power of expression of the theory, solve its counter-intuitive behavior and implement a way to represent, in numeric terms, the uncertainty that comes from unkowledge and conflict among the bodies of evidence.

1.1 Justification

The "Uncertainty concept" has been being one of the most elaborated scientific concepts in recent times [26]. In general, uncertainty in a "problem-situation" emerges whenever the information pertinent to the situation is deficient in some aspect. This deficiency can be caused by a piece of information which is incomplete, imprecise, contradictory, vague, non-reliable, fragmented, or deficient in some other way, giving origin to several kinds of uncertainties [25]. These kinds of uncertainties can be classified in two large groups, objective uncertainty and subjective uncertainty, giving origin to what is known as the "dual nature of uncertainty" defined by Helton in 1997 [23]. However, just recently the scientific and engineering community has started to recognize the usefulness of establishing definitions and treatment models for different kinds of uncertainties [21]. This has been motivated by the extensive development of mathematical analysis since the end of the nineteenth century (by Cantor, Lebesgue, Kolmogorov and De Finneti, just to list some). The applicability of this in every-day situations has been facilitated greatly by the extraordinary computational power easily available today. As soon as systems were able to work with complex analysis, it became easy to discover the limitations of the Bayesian Theory in representing and treating the whole range of uncertainties. A primary motivation to study reasoning subjected to uncertainty is to be able to reach decisions when facing otherwise non-conclusive evidence [54].

The dual nature of uncertainty has been assigned to a concept named "Belief", different from the concept of "Probability", that is used more in the sense of traditional objective (the frequency approach) and subjective probabilities. "Objective probability" (a frequentist approach to probability) can be understood as the probabilistic knowledge obtained on the basis of the relative frequency of the occurrence of events in a long sequence of independent experiments; and "subjective probability" (or Bayesian Probability) as referring to the alteration or the conditioning of a previous probability measure, in function of new evidence or observation.

On the other hand, "Belief" relates to conviction, whether or not this conviction is supported by the concepts of traditional probability. Belief is not necessarily connected to decision-making or betting contexts. It is a cognitive process *per se* [50]. Belief aims to model and quantify both objective and subjective certainties, induced in us by the evidence [50].

Some criteria to support Belief are [53]:

- Faith: a hypothesis is believed because the person who established it is dependable.

- Reasonability: the hypothesis is accepted if it agrees with previously established beliefs.

- Success of Prediction: the hypothesis is believed when it is able to foresee the behavior of events not yet observed.

The Dempster-Shafer Theory is able to formalize these two different kinds of uncertainty simultaneously [25], while the Bayesian Theory models naturally only the objective uncertainty. Because of this the Dempster-Shafer Theory was chosen as the basis for this work.

1.2   Motivation

There are important reasons, practical and theoretical, to study the representation, combination and comparison of beliefs. Humans frequently reason in both objective and subjective terms (understanding this as the quantitative and qualitative aspects of reasoning) [66], thus, the representation of reasoning in the form of beliefs seems to be a more realistic model of uncertainty, particularly when the available information is limited[1]. As there exist situations that demand actions to be taken based not only on the available knowledge, but also on what is recognized as not known, there is a need for the development of non-classic logic, that is, a logic conceived and put into practice to capture particular kinds of reasoning, such as reasoning about beliefs, probabilistic reasoning or default reasoning [31]. These types of logic are not conceived because the classical paradigm supplies wrong answers, but because certain questions cannot be expressed easily, naturally or efficiently (from the computational point of view) [31]. These types of logic result in a meta-theoretical extension of the power of expression of classical logic, since it is not able to talk about ignorance, since reasoning about ignorance results in an odd consequence, non-monotonicity[2] [31], and classical logic is monotonic by principle. Additionally, if belief functions are thought of as generalizations of probability functions (as will be shown below), then the understanding of the mathematical behavior of the beliefs becomes as important as the probability study itself.

Considering these aspects, the motivation for the utilization of the Dempster-Shafer Theory and its extension presented in this work, in the beliefs modeling, arose not just from the technical characteristics of the theory, but also from its wide range of practical applications, mainly in the last 10 years [42], indicating a high degree of experimentation, maturity, and relevance. Practical applications of the Dempster-Shafer Theory can be found in all the sciences, from human to exact. Some examples include image processing, voice recognition, specialist systems, knowledge-based systems, robotics, decision support systems [48], fault

---

[1]Paraphrasing George E.P.Box: "All models are wrong, but some of them are useful" [6].

[2]A property of some non-standard logics. It implies that if there is a change in opinions, or if more is learned, it is possible to arrive at a conclusion that something that was thought as true before is not true anymore. Thus, any logic that must deals with practical subjects should model non-monotonicity [31].

diagnosis, object recognition, biomedical engineering, autonomous vehicles navigation, climatology, simulation, and target identification. Sentz's and Ferson excellent work [42], supplies 148 references for applications of the theory, classified by segment, such as cartography, classification, decision-making, specialist systems, fault detection, medical applications, sensor fusion, risk analysis and reliability, and robotics.

Among the technical characteristics which motivated us to choose the Dempster-Shafer Theory, we can list [42]:

- The relatively high degree of theoretical development.

- The close relations between it and Probability and Set Theories.

- The versatility in the representation and combination of different kinds of evidence obtained from multiple sources (including mixtures of objective and subjective evidence).

2 State of the Art and Related Works

The field of imprecise probability is characterized by closely-related concepts. Thus, if the boundaries of the concepts are not uniform and clearly marked, it is possible to achieve dubious or even mistaken interpretations of the matters presented by this work. Therefore the first part of this chapter will conceptualize a point widely used in this book, the dual nature of uncertainty. This concept is based on Helton's definition [23], which makes the taxonomy of it into two major groups, "Objective Uncertainty" and "Subjective Uncertainty".

Since objective uncertainty has already been extensively explored in works on classic probability, closer attention has been given to subjective uncertainty which is treated by means of imprecise probability and is the subject of this work, which extends one of the formal models that deals with it, the Mathematical Theory of Evidence (or Dempster-Shafer Theory). Section 2.2.1 illustrates the philosophical basis of subjective uncertainty, and Section 2.2.2 its utility in solving questions not naturally modeled by classic probability or logic.

The next section, 2.3, "Models of Knowledge Representation and Combination", confirms the need to consider subjective uncertainty in knowledge representation modeling, exemplifying several imperfections that information can be subject to, along with formal models to treat these imperfections.

The final section, "Probability Interpretations", Section 2.4, presents several possible interpretations of probability, both formal and conceptual. The limitations of classic probability are shown, giving rise to the concept of "belief" (upon which the Theory of Evidence is developed) and again making clear the role of subjective uncertainty in the modeling of probability interpretations.

2.1   Objective Uncertainty

Objective Uncertainty corresponds to the "variability" that emerges from the stochastic characteristic of an environment, non-homogeneity of the materials, time drifts, space variations, or other kinds of differences among components or individuals. This variability is also known as "Type I Uncertainty", "Type A", "Stochastic", or "Aleatory", emphasizing its relationship to the random aspects of games of chance. Another term attributed to it is "Irreducible Uncertainty", since, at least in principle, it cannot be reduced through additional investigation (although it can be better characterized) [21], [42].

2.2   Subjective Uncertainty

Subjective Uncertainty is the uncertainty that comes from scientific ignorance, uncertainty in measurement, impossibility of confirmation or observation, censorship, or other knowledge deficiency. It is also known as "Uncertainty Type II", "Type B", "Epistemic Uncertainty", "Ignorance", or "Reducible Uncertainty", since it, *a priori*, is able to be reduced through additional empiric efforts [21], [42].

Objective or aleatory uncertainty already has well clarified origins and concepts from the origins and concepts of classic probability[1] itself, being the uncertainty that usually comes to mind when we think in contingencies. Thus, this section will limit itself to the treatment of subjective or epistemic uncertainty.

2.2.1   Subjective Uncertainty and Philosophy

The English philosopher John Locke (1632-1704) considered that the called "practical truths" are, in the best of hypotheses[2], "probabilities, approaching authenticity". The knowledge theory of Locke had a materialistic basis; he did not doubt the objective reality of the world that surrounded him and considered probable knowledge as the complement of nature

---

[1]By "classic probability" (or "the classic concept of probability") this work will be referring to the probability described by the Axioms of Kolmogorov, that is, the statistical probability concept (or objective probability) and the concept of Bayesian Theory [1] (or subjective probability).

[2]In this work, hypotheses must be understood as all attempts to describe reality. Hypotheses are built by attributing "variables" to "cases". The "case" is the entity which the hypothesis regards. The "variable" is the characteristic, treatment or attribute that the hypothesis imputes to the case [53].

authentic cognition. Another English philosopher, John Stuart Mill (1806-1873) developed his theory of the hypothetical nature of human knowledge upon a different philosophical premise. Mill was a positivist. He believed that the only source of knowledge is experience, conceiving this as the sum of all that has been experienced by the individual. In fact, Mill doubted the existence of the objective world, since he did not consider matter as objective reality independent of man and existing previous to the sensory perception of itself; he considered matter a continuous possibility of sensations. According to Mill, all of our knowledge is hypothetical and the true nature of the phenomena is beyond our knowledge. V. Kraft a contemporary thinker representative of the neo positivism movement, member of University of Viena and of the Austrian Academy, declared that once the material world that resides outside human consciousness becomes thought of as a hypothesis, then declarations about it can only be considered probabilities rather than truths. [56].

If one limits oneself only to the mathematical formulations, what is called today "belief based reasoning" can be found at the end of 17th century, well before the development of the Bayesian ideas. In 1689, George Hooper supplied the rules for combining testimonies, which can be considered special cases of Dempster's Combination Rule. Other similar rules were formulated by Jacob Bernoulli, in his work "Ars Conjectandi" in 1713, and by Johann-Heinrich Lambert in his "Neues Organon" in 1764 [46].

But where in these hypotheses does doubt come in?

Stephens answers that question very well [53]:

1. An hypothesis can be a forecast: before the predicted instant it represents a source of doubt, whose magnitude depends on a past rate of successful predictions.

2. An hypothesis can be a generalization beyond well-known cases: similar to the forecast problem, it corresponds to the projection of a tendency.

3. An hypothesis can cite an inferred variable: the inference itself, by principle, is a source of doubt.

4. An hypothesis can surmise a cause-and-effect relation: that is, the citation that one variable is the cause of or dependent on other. As it involves correlation between variables, it is subject to three possible explanations, excluding mere coincidence:

   - "A" is the cause of "B".

    - "*B*" is the cause of "*A*".

    - Some other variable or group of variables, "*C*", is the cause of both "*A*" and "*B*", making the covariance of both.

5. An hypothesis can cite a necessary cause: "*A*" can be the cause of "*B*", as the evidence indicate, but also, "*B*" can always be the result of "*A*", or "*A*" is previously necessary to occur "*B*".

6. An hypothesis can be a composed hypothesis: for the hypothesis be entirely correct, all the partial hypotheses must also be correct, increasing the sources of doubt.

7. An hypothesis may manifest doubt owing to the degree of relevance of the evidence supporting it: for example, the doubt coming from the relation between the sample size and the target population of the hypothesis.

8. An hypothesis can present spuriously favorable evidence, eg: biased evidence caused by a sampling mistake.

It must be kept in mind, that "uncertainty" does not always represent negative aspects. In some cases, in fact, it could even be considered a strategic resource, since when an appropriate quantity of uncertainty is allowed in working with certain problems, their associated computational complexity can frequently be substantially reduced [25].

2.2.2 Why cannot everything be solved using standard logic and classic probability?

Standard logic does not have the expressional capabilities to represent uncertainty. For example: assuming that we have two propositions "*A*" and "*B*", both of which can be true or false, the goal of the mathematical standard logic is to determine when derivative expressions, $A \wedge B$, are true or false. However, suppose that because of incomplete knowledge, causing "uncertainty", the specialist does not know whether the propositions A and B are actually true or false, but he can specify the probability of A and of B to be true. At this point, it becomes clear that the goal of the non-standard types of logic (like the probabilistic one) is to evaluate the probability of the expression be true [34]. However, in the classical theory of probability, if we know that "*A*" or "*B*" will definitely happen, but we do not know the probability of A and B individually, we still have to attribute numeric values to them. For example, we would have (using the usual interpretation of classical probability, Laplace's "Insufficient Reason Principle") to say that A and B would have probability of 0.5 each. In a belief theory such as the Dempster-Shafer, we can attribute all our belief to the set $\{A, B\}$, which is different from attributing a weight of 0.5 to each element [31].

The mere existence of uncertainty in a problem does not mean that the classic probability theories are useful. Since the ideal setting for traditional probability involves frequencies, probability is easier to apply when significant frequencies are available. Many problems of subjective judgment are not statistical inference problems, because there is neither a sampling, nor a well-defined population. Nevertheless, statistical evidence representations based on belief functions (which are able to represent the subjective uncertainty) can be useful when combining statistical and non-statistical evidence [46].

With the increase in computational power the implementation of advanced information systems has become possible, endowed with reasoning capabilities that try to imitate nuances of human reasoning, and with that, characteristics such as "uncertainty" became important, since these systems usually use some form of obtaining inferences from knowledge domains where knowledge itself and its implications are uncertain. Inherent among the characteristics of systems that treat uncertain probabilities are their abilities to solve contradictions and to process the relative precision of the inputs for the determination of the resultant probabilities [34], two attributes that are not naturally modeled by traditional probability.

If you take as an example the elaboration of a design methodology, whatever be its application domain, the purpose will be to systematize and to optimize the project space exploration. This exploration needs a modeling step or systematization of the project variables and a mapping step for these variables, whether this mapping is of components, orientations, standards, or platforms. The project variables, arising from the designer inputs, consist of the representation of functional and non-functional requisites, which move over the same characteristics many times in a conflicting way and with different degrees of certainty. Some examples:

- A high processing speed conflicts with the low consumption requisites, battery utilization and reduced size (due to the needs of dissipation).

- A high degree of predictability and reliability in processing requires processors with special treatment or military standards, and in any case eliminates the possibilities of choosing the most up-to-date PC processors.

Consequently, the project of a system with any interesting degree of complexity, far from being a Cartesian and deterministic process, is a process of continuous commitments between different requisites, often conflicting and based on resources about which the probabilities of their uses, *a priori* cannot be affirmed. The designer or design team itself, as much as it is experienced and prepared, does not have the means to determine at the very beginning of a project the best commitment between all the variables, since there are open questions (questions of extreme complexity if we aggregate them to the non-functional requisites), the number of involved variables is very large and their interrelation is usually non-trivial and unpredictable to some point.

2.3 Knowledge Representation and Combination Models

Taking as an example knowledge-based systems, if it were always possible to get perfect information from users and if the knowledge-base had a perfect modeling with respect to the information it provided (always providing precise answers for each question, without uncertainties and conflicts between its records), the establishment of a mapping between the data inputs of users, and the correct answers in the knowledge-base would not be a difficult task. However, practical reality shows that imperfect information will always get into the data-base and an imperfect knowledge-base will always exist, or else we would have just limited application systems, since they could only be used by specialist users who had total certainty about their applied inputs and whose knowledge-base would have been elaborated by specialists able to supply precise and non-conflicting answers for each possible question.

Example 2.1 *Example of reasoning based on ignorance: if a database does not specify that a flight stops in a given city, we assume that this flight does not stop in this city (considering the "Closed World Principle") [31].*

The uncertainty can be derived from several sources [5], [53], for example from the partial reliability of the information, the inherent imprecision of the language in which the information is expressed (or of the device used to get it), the incompleteness of the information, and the aggregation or the summing up of the information which comes from multiple sources.

The literature in the area sometimes treats any information imperfection as "uncertainty", although several authors like [45], [69], [5], find this term very limited, since the so-called "uncertainty management" also encompasses imperfections like imprecision, conflict of evidence, and partial ignorance. The following example illustrates several kinds of information imperfection [4]:

Example 2.2 *Imagine that we wish to know John's score in a certain discipline, and to do this we ask several people about this score, obtaining the answers (in reality, John's score was 7):*

- *Perfect information: "John's score was 7".*

- *Imperfect information: "John got a score of 5 in this discipline".*

- *Imprecise information: "John's score was between 6 and 9".*

- *Uncertain information: "I think that John's score was 7, but I am not sure".*

- *Vague information: "John's score was around 7".*

- *Probabilistic information: "It is probable that John got a score of 7".*

- *Possibilistic information: "It is possible that John got a 7".*

- *Inconsistent information: "Maria said that John obtained score of 6, but Carlos said that his score was 10".*

- *Incomplete information: "I do not know John's score, but the class got an average grade of 6 in this discipline".*

- *Total ignorance: "I don't have the slightest idea what John's score was.".*

- *Partial ignorance: "John earned a 6 in the first exam of this discipline, I do not know his second exam grade".*

Therefore, it is observed that the quality of the obtained information can vary widely, whether because of precision, partiality, conflict or the other reasons suggested above. Human beings continually manage these different degrees of information quality, utilizing *ad hoc* models created to deal with and manipulate the various kinds of information imperfection. However, when we think about the implementation of artificial reasoning systems, the utilization of *ad hoc* models introduces significant deficiencies, based in the fact that these models are neither subsidized by a well supported theory, nor do they have the support of a well defined semantics system.

The literature already suggests formal models for the treatment of each one of these kinds of imperfection [4], although there are efforts toward a unified model [22], [62]; Table 1 summarizes these, in a non-exhaustive, non-determinant way[3].

The applicability of some of these models can be shown for other kinds of uncertainties not the listed in the table[4], in particular, it is possible to show that the Theories of Possibilities, Probabilities, and Fuzzy Sets can be modeled as particular cases of the Theory of Evidence, vide Section 3.2.

Formal Model	Probabilistic	Imprecise or Vague	Possibilistic	Uncertain	Inconsistent	Incomplete	Imperfect	References
Probability Theory	XXX			XXX				
Theory of Evidence	XXX			XXX				[45]
Fuzzy Sets		XXX						[68][18]
Rough Sets		XXX						[35]
Reference Classes		XXX						[28]
Possibilities Theory			XXX	XXX				[69] [19]
Para-consistent Logic					XXX			[2]
4 Valued Logic					XXX			[2]
Default Logic						XXX		[38]
Circumscription Logic						XXX		[32]
Local Propagation							XXX	[36] [40]

Table 1: Formal models to deal with different kinds of information imperfection

As it can be deduced from Table 1, either the Theory of Evidence (or Dempster-Shafer Theory) or the Probability Theory (or Bayesian Theory), can be used to solve the same class of problems, being especially indicated to work with uncertain or probabilistic information. In fact, as already has been demonstrated since Shafer's seminal work presenting the Theory of Evidence [45], [4], a Bayesian probability function is a particular case of a Belief function, and the Bayes's rule of conditioning, a special case of the Dempster's rule, implying that the Dempster-Shafer Theory contains the Bayesian Theory.

---

[3]There are other models not cited in this work, for example, the models based on aggregation of opinion [41], the "Bag of Marbles" [59], and the "coherent upper and lower previsions" [61].

[4]The problem of knowledge eduction is outside the scope of this work, but it represents a whole field of open questions.

2.4 Probability Interpretations

2.4.1 Formal Interpretations

Two works published in the 30's, the Axioms of Kolmogorov, related to classic probability, and those of De Finetti, regarding the qualitative aspects of probability are considered from a formal, axiomatic, point of view to be the most acceptable.

One can utilize the ideas of Kolmogorov and define "probability" as a normalized and denumerable additive measure, defined upon a σ-algebra[5] of subsets in a abstract space. For a finite space probability can be expressed as numbers between 0 and 1 such that if two events[6] can not occur simultaneously, the probability of either of them occurring $(P(A \cup B))$ is the sum of the probability of the first one with that of the second [37].

The Axioms of Kolmogorov consider a measure space Ω, a σ-algebra $\mathcal{F}$ of Ω subsets and a measure P, related to themselves by the axioms:

I. $\forall A \in \mathcal{F}, \exists P(A) \geq 0$

II. $P(\Omega) = 1$

III. σ-additivity: $\forall A_i, A_j \in \mathcal{F}$ as such $\forall i, j \ \ A_i \cap A_j \neq \varnothing$, it follows that

$$P \left(\bigcup_{i=0}^{\infty} A_i \right) = \sum_{i=0}^{\infty} P(A_i)$$

The previous axiom implies in the so-called "additivity":

$$\text{if } A \cap B = \varnothing, \ P(A \cup B) = P(A) + P(B)$$

From these axioms one can demonstrate the following results [15] (vide [14] for the demonstrations):

---

[5]σ-algebra: set of events of a sample space, defined by the following properties:

- For each event that belongs to the algebra, its complement (in relation to the sample space) also belongs to the algebra;

- The union of all events that belong to the algebra also belongs to the algebra.

[6]In this work the Moivrean concept of event is used; that is, a subset of the sample space (to know more about Moivrean Events vide [47]). In an analogous way, two events, $\mathcal{F}$ and $\mathcal{G}$, are considered independent if there is no situation able to influence both. In the same way, two variables are considered independent when the situations that influence one of them do not influence the other.

1. $\forall A \in \mathcal{F}, P(A) \leq 1$, that is, the probability of any event is always less than or equal to 1.

2. $P(\varnothing) = 0$, implying that the probability of the impossible event is equal to 0.

3. $\forall A \subset B, P(A) \leq P(B)$, that is, if a set is contained in another (meaning that the occurrence of the first event implies the occurrence of the second one), the probability of the first always will be less than or equal to the probability of the second.

4. $P(A \cup B) = P(A) + P(B) - P(A \cap B)$, meaning that the probability of the union of two events is equal to the probability of the first plus the probability of the second minus the probability of the occurrence of both simultaneously.

5. $\forall A, B \in \Omega, P(A \cup B) \leq P(A) + P(B)$, which is a consequence of the previous result, showing that the probability of any two events happen, together or separately, never is larger than the sum of the probability of occurrence of each one.

Regarding the Axioms of De Finetti, events such "A", "B" and "C" are considered, where:

- by $A \geq B$ can be read "event A is at least as probable as event B";

- the events A and B are equally probable, that is "$A \cong B$", if $A \geq B$ and $B \geq A$;

- if $A \geq B$ but not $A \cong B$, it is written "$A > B$";

- and by "$A + B$" one denotes an event composed of "A or B".

These events are related by the "Axioms of the Qualitative Probability", based on the notion of "coherent bets" [37], [63]:

I. $A \geq B$ or $B \geq A$ to any events A and B.

II. $A > C > B$ when A is sure, B impossible and C none of these.

III. $A \geq B$ and $B \geq C$ implies $A \geq C$.

IV. If A and B are both incompatible with C, $A + C \geq B + C$ if, and only if, $A \geq B$. Specifically, $A \cong B$ if, and only if, $A + C \cong B + C$.

De Finetti defends the idea that these axioms are the basis of subjective probability, considering probability as a primitive concept of human behavior under uncertainty, possessing a basically qualitative nature, although being able to be measured numerically through bet rates that the one subject would be wiling to accept for the occurrence of an uncertain event.

It must be stressed that the formal definitions of the classic probability do not explicitly describe the special situation on which it is based, such as the existence of a well characterized population and a long sequence of repetitions of a given situation [44]. When these special conditions are not present, the classic probability concepts and definitions are not enough or are not valid to model all the possible situations, and in particular, the ones that introduce subjective uncertainty [44]. The "probability" turns out to be insufficient if considered only in its classic sense, because it just models the objective uncertainty and its concept no more can be clearly identified with the traditional "expected frequency situation" [44]. At this point the concept of "belief" emerges, that can be understood as a "general probability" where both the objective and subjective uncertainties are modeled. In the following section, the meaning of an attribution of probability to an event will be explored from a conceptual point of view.

2.4.2 Conceptual Interpretations

Although intuitively all of us have a notion of what "probability" is, since, without even thinking about it, probability is a characteristic of phenomena that we have been observing and experimenting all our life, although, whenever we try to express its concept in words, we encounter a problem of multiple meanings for some words and multiple words with the same meaning, resulting in widely diverse conceptualizations along the centuries of the history of probability. So, in this section we are entering in an environment which is at the same time subtle and controversial, but of extreme importance to the understanding of the need to adopt a new model of belief (or probability in a wide, generic, sense) treatment, different from the ones traditionally adopted.

The probability ideas derive from a history about the relation between an observer and the world, nature [47]. Across the centuries[7] several attempts at probability conceptualization have been made. Just to illustrate we will cite one from Laplace and another from

---

[7]In this work, we do not intend to write an essay about the history of probability. This already has been done very well in several works like [13] and [37].

Bernoulli (considered the father of the mathematical probability). Laplace said that "all the probabilities are relative to the knowledge and to the ignorance, and the completeness of the knowledge in the limit simply eliminates them" [45]; while for Jacob Bernoulli, contingency, and consequently probability, are subjective, once "all things that exist are necessary and certain in and of themselves[8]. Things can be contingent and uncertain or partially certain, only relative to our knowledge"; being the probability the degree of subjective certainty [47]. According to Shafer, this degree, describes the individual's knowledge and ability to foresee. "When I say it is contingent whether a coin will fall heads, I am acknowledging that I cannot calculate how it will fall. When I give a probability for its falling heads, I am expressing precisely my limited ability to predict how it will fall" [47].

Even with the help of these concepts, it is possible that the exact meaning of the following statements may still not be clear:

"The probability of event A is 75%."

What does this mean?

Does it have to do with the frequency of the occurrence of the event?

Does it represent the degree of certainty that we believe that it will occur?

Does it has to do with someones's degree of belief that the event will occur?

Or does it simply mean that someone finds it very probable that this event will happen?

To try to answer that question, we are going to give an explanation based mostly on the works of Shafer [44], [46] and Paass [34].

According to Shafer in [44], degrees of belief, frequencies, and degrees of evidential support are all subjected to the same mathematical laws of probability, as well as the axioms and definitions of Kolmogorov. Thus, in principle this would allow for three interpretations, to which we add a forth one, symmetry:

1. Symmetry:
   The situation is symmetric enough for one be able to estimate the rate of occurrence for a given result, implying the probability of occurrence of this event. As examples would be the occurrence of a given result in the casting of balanced dice or coins, where

---

[8]The concept of "Chance".

the symmetries involved in the situation enabled the prediction of the occurrence of a given number, A, in the casting of the dice, as a probability ratio of 1/6; or, in the case of a coin, one chance out of two of it coming up heads.

2. Bet Rate:

   Interpretation of a fair rate at a wager. When a person says that he or she believes in A with a probability of 50%, we assume that he or she is able to bet on A at a rate of 2:1 (for each "1" bet she will pay "2"), or worse (1:1, 0.9:1, etc.).

3. Frequency:

   The probability of an event is the frequency, after many repetitions, with which the event occurs in a given experimental arrangement or in a particular population. This frequency is a fact about the experiment or about the population, and this fact is independent of any personal belief.

4. Support:

   In this interpretation, probability is the rational degree of belief. The probability of an event A is the degree to which we must believe that A will happen, or the degree to which the evidence supports the occurrence of A.

   The question is, can a precise numeric degree be assigned to this support?

   Contemporary proponents of this interpretation consider that it is difficult to measure the degrees of support in an numerically precise[9] way, but they maintain that the evidence provides support for the belief, and depending on the degrees of conflict and ignorance, this support can be of a more qualitative than quantitative nature.

A way to analyze these four interpretations is to consider that the "betting rate" and the "symmetry" correspond to the usual probability concept, the "frequency" to the statistical probability concept, and the "support" to the more generic concept of belief, where both objective and subjective uncertainties can be modeled. But, it is important to keep in mind that these interpretations are not necessarily exclusive, since the third interpretation, "Support", can include the two firsts interpretations, since the probabilities, in the traditional sense, of a certain aleatory experiment may or may not coincide with our degrees of belief about the result of the experiment. If we are previously aware of the probabilities, we

---

[9]Considering the concept of "precision" adopted by Lins and Campello de Souza in [30].

certainly must take them into account as our degrees of belief, but if we are not aware of them, it would be an extraordinary coincidence for our degrees of belief to be equal to them [45].

Now let us move on to the concept of "support" regarding the modeling of the subjective uncertainty (or imprecise probability):

Suppose that $\mathcal{B}$ represents "the patient has hypertension", and that the patient, John, consulted a specialist, who after some investigation affirmed that $P(\mathcal{B}) = 0.3$ [34].

What is the meaning of this?

The usual concept of probability involves a long sequence of repetitions of a particular situation. For example, to say that a coin possesses a 50% probability of landing heads up means that in a long sequence of independent plays the head will land heads up half of the times. However, this frequency concept is not ideal to work with in situations such as our patient John, since it is not possible to clone him, with exactly equal life histories, in order to count the relative frequency of all those Johns with hypertension [34]. In this case, we have a "subjective probability", represented by a probability in the form of a belief that the proposition may be true. In this context, the proposition can be characterized as a clear declaration that can be true or false. Instead of trying to determine the relative frequency of the event, the subjective probability was determined through anamnesis.

This behavior can be generalized as a form of uncertain knowledge about the probabilities [34]. Let us assume, for example, that a number of specialists supplied numbers expressing their subjective probabilities concerning facts or rules of interest to the decision maker. As these probabilities possibly have different probability distributions according to the limited amount of information and individual experience of each specialist, his or her judgment is uncertain, and could also be erroneous or conflicting to some degree. This does not mean that the specialists have violated the laws of rational decision-making, since these differences can also be attributed to different knowledge backgrounds and degrees of authority on the topic at hand.

Every time we talk about the degree of support that a particular body of evidence provides for a proposition, or degree of belief that an individual attributes to a proposition, there is an act of judgment. This does not necessarily mean that there is an objective relation between a particular body of evidence and a given proposition, which determines a precise degree of numeric support. Neither it is expected that some individual mental state concerning a proposition can be described precisely with a corresponding real number representing its degree of support. Nor can the determination of such numbers even be always expected. But, it is assumed that an individual is able to make a judgment. Having verified perceptions and understandings (which constitute an body of evidence), sometimes vague and other times confuse, our subject announces a number that represents the degree with which he has judged that the evidence supports a given preposition, and consequently, the degree of belief that he wishes to attribute to this proposition [45].

On the other hand, assuming that a specialist would have specified his personal degree of belief in the truth of the propositions under consideration, it is clear that all this numeric degree of belief would not constitute a measure of probability (in the classic sense of probability), P. It can be shown, however, that a probability measure would result if the belief specifications obeyed a series of axioms that mimicked rational behavior. Such axioms, for example, might contain the postulates for the behavior of a rational gambler, which would specify his preferences among available bets; the attractiveness of these axioms, and consequently the probability laws derived from them, comes from the fact that when someone violates them using a different scalar measure of uncertainty, the outcome is subject to a demonstrable, and thus irrational, loss [34].

Theories based on belief functions provide a method for the use of mathematical probability in subjective judgment. Usually, these theories consist of a generalization of the Bayesian theory of subjective probability, where the degrees of belief either can or cannot have the mathematical properties of classic probability [46].

2.4.3 Interpretations Concerning Uncertainties

From the middle of the 17th century, when the concept of numeric probability emerged, until the middle of the 20th century, uncertainty was treated only in terms of the Bayesian Theory. This connection was considered a de facto law for three centuries. Questions about the connection arose through the development of several mathematical theories that are more general, or totally distinct, from the Bayesian Theory. A form of generalizing Probability Theory is to allow the use of imprecise probabilities, that is, to consider the subjective uncertainties. The formalization of these imprecise probabilities can be accomplished in different ways, one of which is the Dempster-Shafer Theory. In the case of the Dempster-Shafer Theory, imprecise probabilities are represented by means of two measures, one of which is super-additive (the "Plausibility Function") and other, sub-additive (the "Belief Function"). These measures replace the classical concept of the additive measure in probability theory [25][10].

Belief functions are well equipped to represent ignorance, once its rules allows one to express a frank agnosticism through the attribution of a low degree of belief to both a proposition and its denial. In fact, they could allow the attribution of a zero degree of belief to all subsets of a frame of discernment, resulting in the called "vacuous belief function".

Bayesian Theory, on the other hand, does not have the resources to manage the representation of ignorance so naturally, since it is not able to distinguish lack of belief in an event from the lack of knowledge regarding it.

In both theories there is the need to provide a method to modify a previously attributed probability to contemplate the knowledge coming from new bodies of evidence. In the case of the Dempster-Shafer Theory, the method used is Dempster's Rule of Combination[11], and in the case of the Bayesian Theory, Bayes' Conditioning Rule. Dempster's Rule deals symmetrically with the evidence, no matter what order the evidence is in, or which of the bodies of evidence was known first. In the case of Bayes' Conditioning Rule, new evidence is represented as a proposition and this proposition is used to condition our previous belief, resulting in an asymmetry in dealing with both the new and the previous evidence (if you modify the older with the new, the numeric results of the Bayes' conditioning rule will, in general, change also). Furthermore,it assumes that the exact effect of new evidence is to establish a proposition with certainty.

Since Dempster's Rule does not compel one to express the evidence as a certainty, it allows for the building of probable reasoning descriptions that are more moderate than the purely Bayesians descriptions, and at same time more faithful to the epistemic reasoning mechanisms. Statistical inference students are accustomed to the idea that "chance" cannot be evaluated with less than infinite evidence. In order to establish a value between 0 and 1 as the chance resulting from an aleatory process, the results of an infinite sequence of a

---

[10]Besides this point, the remaining ideas of this subsection are strongly based on the Shafer's works, in particular [45].

[11]That will be changed later, in this work, with our extension to the Dempster-Shafer Theory

process of independent executions must be obtained, and the proportion of these executions must converge to values not equal to 0 or 1 as this sequence continues. An example of "infinite" would be a precisely balanced and non-obtainable evidence[12]. Therefore it follows that "chances" are hypothetical rather than practical, and not able to be directly translated into degrees of support for events. They can often be incorporated indirectly, however, into a frame of discernment by associating different beliefs with events that have different possibilities of occurrence; and this serves to facilitate the evaluation of these events.

In a sense, a Bayesian belief function indicates an infinite quantity of evidence in favor of each possibility to which it makes attributions. If there were a reasonable convention for the *a priori* establishment of probability distributions, a convention that could be applied without the need for arbitrary choices that so strongly affect the final results, we would have a strong case for the utilization of the Bayesian Theory instead of the Dempster-Shafer Theory.

However, it is evident that the difference between the methods is primarily in how they deal with conflict and and how they manage the ignorance coming from the observations. When the bodies of knowledge are combined at an epistemic level, as they are when the Dempster-Shafer Theory is used, the information concerning conflict and ignorance is preserved [43]. But combining only at the aleatory level, as in the Bayesian Theory, forces our final belief function to suppress every bit of conflict or ignorance in the bodies of knowledge [43].

The next chapter will explain the Dempster-Shafer Theory in more detail.

2.5 Conclusions

There are two aspects of uncertainty: objective and subjective. Both need to be modeled if the purpose is to provide a broad modeling of the knowledge coming from imperfect sources of information.

Objective uncertainty is treated by classic probability, but subjective uncertainty exhibits several difficulties for modeling using this approach.

---

[12]As Courant said in [12], we must explicitly emphasize that the symbol ∞ does not denote a number with which we can calculate as we would with any other number; equations or statements expressing a quantity that be or become infinity never have the same sense as an equation that deals only with defined quantities. Although this mode of expression and the use of the symbol "∞" are extremely convenient...

Because of the limitations of the classic probability in the modeling of subjective uncertainty, the concept of "belief" is used to indicate the models able to deal with both kinds of uncertainty. The modeling of these two kinds is necessary to deal with all interpretations of the concept of probability.

3   Dempster-Shafer Theory

The Dempster-Shafer Theory, also known as the "Theory of Evidence" is a simple and versatile[1] formalism, particularly interesting for providing methods to combine evidence from different sources, without the need to know, *a priori*, their probability distributions. It was formally introduced in 1976, through a work by Shafer [45], based on an extension of Dempster's works [16].

Different from the Bayesian Theory, the Theory of Evidence does not necessarily require previous knowledge of the distribution of the probabilities since it is able to assign probability values to sets of possibilities, rather than just to simple events. Another differential is that there is no need to divide all the probability among the events, once the probability non-attributed to any event is, in reality, attributed to the environment, and not to the remaining events. This latter characteristic makes it possible to model a situation where the user has some belief in a given hypothesis but is aware of the possibility of other options which might be the best solution. Since he does not know which option will provide the best solution, rather than having to divide his non-assigned belief among the remaining hypotheses, he can use the environment feature of the Theory of Evidence to avoid this. These two differentials allow a more precise modeling of the natural process of reasoning based on the accumulation of evidence, making the theory progressively more popular.

The methods for the combining of evidence that arises from different sources are called "combination rules", Dempster's Rule being the *de facto* method [17], although other combination rules exist they differ from the Dempster one basically as to what constitutes normalization [24], [4]. The procedures adopted for all the combination rules are independent of the order of evidence.

---

[1]In the sense of encompassing several other theories as particular cases of it, like the Bayesian Theory, the Possibilities Theory, the Fuzzy Set Theory and the Certain Factor Model, just to cite some of them (vide the final of this section for more details).

The Dempster-Shafer Theory allows for the expression of partial beliefs when it would be impossible or non-practical to assign probability distributions confidently. Therefore, attributing belief to events for which the precise assignment of probability would be confusing becomes possible [54], providing a comprehensive and convenient means for the handling of several problems, among them [21]:

- Imprecise distribution specifications.

- Barely known or unknown dependencies.

- Non-disposable uncertainty in measurements.

- Detection problems and other kinds of censorship in measurements.

- Small sample size.

- Inconsistency in the quality of data input.

- Uncertainty or non-stationary models (non-constant distributions).

Despite these advantages, there are classical problems with the theory, already suggested by Shafer in [45], and afterward explored by several authors, that can affect its practical application. These problems will be treated in more detail in Section 4.1. They consist basically of the absence of an intrinsic way of representation, in the result, of the uncertainty and conflict among the evidence, and of counter-intuitive behaviors of the rules of evidence combination. These problems will be solved by our extension of the Dempster-Shafer Theory, increasing its power of expression and range of applications.

On the other hand, traditional questions, like those referring to the computational feasibility of the Dempster-Shafer Theory were solved by the works of Kong (1986), Shafer and Shenoy (1988), Almond (1988), and Chandru (1990), among others (vide [46] and [11] for more explanations and detailed references). And the conclusion arrived is that if a problem is computationally feasible using a Bayesian solution, it will also be feasible using a solution based on the Dempster-Shafer Theory [71].

3.1 Description of the Dempster-Shafer Theory

3.1.1 Frame of Discernment

The Dempster-Shafer Theory presupposes the existence of a set of primitive, atomic, hypotheses[2] called "environment", "medium", "problem domain", "universe of discourse", or "frame of discernment", represented by Θ. The "frame of discernment" must:

- Be exhaustive, in the sense of being complete, containing all possible primitive (atomic) solutions to a problem or question.

- Have mutually exclusive primitive elements.

When a proposition corresponds to a subset of the frame of discernment, we say that the "frame of discernment discerns this proposition" [45].

Each subset of Θ formed by the disjunction of its elements, can be interpreted as a possible new hypothesis, resulting in 2^{Θ} possible hypotheses. But with respect to a particular domain, not all the subsets may be of interest. Since the elements are mutually exclusive and the environment exhaustive, there can be only one subset with the correct answer.

Example 3.1 *Possible scores of John in his German Course:* $\Theta = 1, 2, 3, 4, 5, 6, 7, 8, 9, 10$ *(vide Picture 1).*

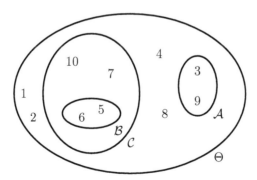

Figure 1: Frame of Discernment of John's possible scores in the Course of German

---

[2]The Dempster-Shafer Theory for the continuous case is explained in [65], [64] and [33].

3.1.2   Mass Function

The Dempster-Shafer Theory uses a basic probability assignment function[3], also known as "probability basic mass", or "mass function", to assign a quantity of belief to the elements in the frame of discernment.

To consider a particular piece of evidence, the mass function, m, assigns to each possible subset of Θ (that is to 2^Θ, the powerset of Θ), inclusive to Θ itself, a number in the interval $[0, 1]$, where 0 means no belief, and 1 represents certainty, in such a way that the sum of all these attributions, including the value attributed to Θ, is 1. This requirement that the sum's total be equal to 1 represents the same thinking as saying that the right answer be contained within the frame of discernment; which is guaranteed, since Θ is exhaustive by definition. In a similar way, 0 must be assigned to the empty set, once it corresponds to the false hypothesis.

The quantity 1 less the sum of the values attributed to the subsets of Θ, that is, a probability not assigned by the evidence to any subset of Θ, but to Θ itself, is named "non-assigned belief", $m(\Theta)$; being in fact assigned to the environment, and not to the denial of the hypotheses that received belief, as it would be in the Bayesian Model. Thus $m(\mathcal{A})$ is a measure of the belief assigned by a given evidence to $\mathcal{A}$, where $\mathcal{A}$ is any element of 2^Θ. It must be stressed that $m(\mathcal{A})$ refers to the belief attributed to $\mathcal{A}$ only, and not to its subsets. Consequently, none-belief is forced by the lack of knowledge of a hypothesis, since the quantities are designated only to those subsets of the environment to which ones wish to assign belief. The quantity $m(\mathcal{A})$ is called "basic probability number of $\mathcal{A}$", being understood as a belief measure that is committed exactly to $\mathcal{A}$.

It is important to notice a significant difference regarding the classic probabilities (the Kolmogorov approach), that associates a number in the interval $[0, 1]$ to each atomic element of the σ-algebra, so that the sum of these attributions results in 1. This is a specialization of the procedure utilized by the Dempster-Shafer Theory.

If we consider the elements of Θ as regions, we can think of our belief portions as semi-mobile probability masses, which can move themselves from region to region, without restriction in several of these regions. So, $m(\mathcal{A})$ measures the probability mass that is confined to

---

[3]One must keep in mind that although the term is "basic probability assignment", it does not refer, in general, to the probability in the classical sense, but in the belief sense [42].

$\mathcal{A}$ but that can move freely to each region contained in $\mathcal{A}$ (considering, obviously, that $\mathcal{A}$ is not a singleton[4]) [45]. This is different from traditional probabilities distribution, which distributes the belief on singletons in the sample space, while this distribution that we call "mass function", attributes belief to subsets of the sample space. The belief attributed to a subset implies in the belief that the correct answer to the proposition is among the elements of that subset, without the need to bind a particular quantity of belief to any of the elements [54].

Summarizing:

$$m : 2^{\Theta} \to [0, 1] \tag{3.1}$$

$$m(\varnothing) = 0 \tag{3.2}$$

$$\sum_{\mathcal{A} \in \Theta} m(\mathcal{A}) = 1 \tag{3.3}$$

A subset $\mathcal{A}$ of a frame of discernment Θ is named "focal element" if $m(\mathcal{A}) > 0$.

Example 3.2 *Mass Function for an evidence of John's score in the German Course*[5] *(vide Figure 1):*

Basic belief mass (m):

$$m_1(\mathcal{A}) = 0.3$$

$$m_1(\mathcal{B}) = 0.25$$

$$m_1(\mathcal{C}) = 0.35$$

$$m_1(\Theta) = 0.1$$

Note that the belief non-assigned to the subsets (whose value is 0.1), is attributed to the environment, and not to the remaining elements.

---

[4]Singletons are the unitary subsets of the sample space, that is, the primitive, atomic, elements of the frame of discernment.

[5]In this work the numeric subscripts denote that the quantity in question comes from a given source, thus, "m_i" indicates the basic belief mass that comes from the source "i".

3.1.3   Body of Evidence

According to the Dempster-Shafer Theory, the information supplied for a given knowledge source regarding the real value of a variable x, (the "evidence" in lay terms), defined in a frame of discernment Θ, is coded in the form of an "evidence body" upon Θ. An evidence body, $\mathcal{EB}$, is characterized by the tuple $(\mathcal{F}, m)$, where $\mathcal{F}$ is a family of subsets of Θ, (that is, $\mathcal{F} \subseteq 2^\Theta$), and m is a mass function.

Example 3.3 *Using the data from Example 3.2:*

$$\mathcal{F}_1 = \{\mathcal{A}, \mathcal{B}, \mathcal{C}, \Theta\}$$
$$\mathcal{EB}_1 = (\mathcal{F}_1, m_1)$$

3.1.4   Belief Function

The "Belief Function", or "Credibility Function", $\mathcal{B}el$, measures to what extent the information given by a source supports the belief in a specified element , $\mathcal{A}$, belonging to $\mathcal{F}$, as the right answer. To get it we must sum to $m(\mathcal{A})$ the values of m for all its own subset, $\mathcal{B}$, of $\mathcal{A}$:

$$\mathcal{B}el : 2^\Theta \rightarrow [0, 1] \tag{3.4}$$
$$\mathcal{B}el(\mathcal{A}) = \sum_{\mathcal{B} \subseteq \mathcal{A}} m(\mathcal{B}) \tag{3.5}$$

To the union of all the focal elements of a belief function we name "core" or "kernel" [45].

Example 3.4 *Still using data from Example 3.2:*

$$\mathcal{B}el(\mathcal{A}) = 0.3$$
$$\mathcal{B}el(\mathcal{B}) = 0.25$$
$$\mathcal{B}el(\mathcal{C}) = 0.6$$
$$\mathcal{B}el(\Theta) = 1$$

Note that the belief in $\mathcal{C}$ is the sum of the belief mass of $\mathcal{B}$ (that is 0.25) with the belief mass of $\mathcal{C}$ (0.35), given that $\mathcal{C}$ contains $\mathcal{B}$, that is: $\mathcal{B}el(\mathcal{C}) = m(\mathcal{B}) + m(\mathcal{C}) = 0,6$ once $\mathcal{B} \subset \mathcal{C}$ and $\mathcal{C} \subseteq \mathcal{C}$. And the belief in Θ corresponds to the sum of all the attributes of the belief masses, always resulting in 1.

Belief functions, in general, should not be interpreted as lower probabilities (using "probabilities" in the traditional sense), because they are not, in general, related to a well-defined reference population, from which we can learn about their frequencies [46]. On the other hand, the additive belief degrees from Bayesian Theory correspond to an intuitive setting where the total belief of an individual is susceptible to division into atomic portions (singletons) and always committed to propositions in such a way that the rest of them are committed necessarily to the denial of these propositions [45].

It is also relevant to cite the relation between mass function and belief function, the Möbius Inversion Formula [60]:

$$m(\mathcal{A}) = \sum_{\mathcal{B} \subseteq \mathcal{A}} (-1)^{|\mathcal{A} - \mathcal{B}|} \mathcal{B}el(\mathcal{B}) \tag{3.6}$$

3.1.5 Plausibility Function

The "Plausibility Function", $\mathcal{P}l$, measures to what extent the information given by a source does not contradict the specified element, $\mathcal{A}$, as the right answer, or in other words, how much we should believe in an given element, $\mathcal{A}$, if all unknown belief is assigned to it, that is, the maximum quantity of belief that can be attributed to $\mathcal{A}$.

$$\mathcal{P}l : 2^{\Theta} \rightarrow [0, 1] \tag{3.7}$$

$$\mathcal{P}l(\mathcal{A}) = \sum_{\mathcal{B} \cap \mathcal{A} \neq \varnothing} m(\mathcal{B}) \tag{3.8}$$

$$\mathcal{B}el(\mathcal{A}) \leq \mathcal{P}l(\mathcal{A}) \mid \mathcal{A} \subseteq \Theta \tag{3.9}$$

$$\mathcal{P}l(\mathcal{A}) = 1 - \mathcal{B}el(\mathcal{A}') \tag{3.10}$$

(where $\mathcal{A}'$ is the complement of $\mathcal{A}$)

Belief and the plausibility functions are not additive.

Neither requires that the sum of its own numeric results be 1 [42].

Example 3.5 *(using the data from Example 3.2)*
For $\mathcal{A}$:

$$\mathcal{A} \cap \mathcal{A} = \mathcal{A} \quad \neq \varnothing \quad m_1(\mathcal{A}) = 0,3$$
$$\mathcal{B} \cap \mathcal{A} = \varnothing$$
$$\mathcal{C} \cap \mathcal{A} = \varnothing$$
$$\Theta \cap \mathcal{A} = \mathcal{A} \quad \neq \varnothing \quad m_1(\Theta) = 0,1$$

$$\mathcal{P}l(\mathcal{A}) = 0,4$$

For $\mathcal{B}$:

$$\mathcal{A} \cap \mathcal{B} = \varnothing$$
$$\mathcal{B} \cap \mathcal{B} = \mathcal{B} \quad \neq \varnothing \quad m_1(\mathcal{B}) = 0,25$$
$$\mathcal{C} \cap \mathcal{B} = \mathcal{B} \quad \neq \varnothing \quad m_1(\mathcal{C}) = 0,35$$
$$\Theta \cap \mathcal{B} = \mathcal{B} \quad \neq \varnothing \quad m_1(\Theta) = 0,1$$

$$\mathcal{P}l(\mathcal{B}) = 0,7$$

And for $\mathcal{C}$:

$$\mathcal{A} \cap \mathcal{C} = \varnothing$$
$$\mathcal{B} \cap \mathcal{C} = \mathcal{B} \quad \neq \varnothing \quad m_1(\mathcal{B}) = 0,25$$
$$\mathcal{C} \cap \mathcal{C} = \mathcal{C} \quad \neq \varnothing \quad m_1(\mathcal{C}) = 0,35$$
$$\Theta \cap \mathcal{C} = \mathcal{C} \quad \neq \varnothing \quad m_1(\Theta) = 0,1$$

$$\mathcal{P}l(\mathcal{C}) = 0,7$$

Plausibility and belief functions must not be interpreted, in general, as unknown classic probability bounds[6]. Although such interpretation seems plausible when we consider a body of evidence on its own, it is not plausible anymore when we consider the combination of conflicting evidence. Furthermore, a probability bounds interpretation is not compatible with Dempster's rule of combination [46]. But it can be shown that if no post-combination normalization is necessary, the belief and plausibility functions proceed to correspond to the traditional probability bounds. That can also occur when normalization exists, but not in a general manner [46].

3.1.6 Belief Interval

Once Pl has indicated how much is possible to believe in a certain hypothesis, and Bel the current belief, and considering that it can be established that $Bel(A) \leq Pl(A)$, it is now possible to express the belief contained in the hypothesis A by an interval, the "belief interval", $I(A)$:

$$I(A) = [Bel(A), Pl(A)] \tag{3.11}$$

This interval expresses the probability range in which is possible to believe in hypothesis A without severe errors; as it is as broad as the uncertainty of the belief in A.

Example 3.6 *(using the data from the Examples 3.4 e 3.5)*

$$I(A) = [0.3 , 0.4]$$

$$I(B) = [0.25 , 0.7]$$

$$I(C) = [0.6 , 0.7]$$

In the Probability Theory, $Bel(A) = Pl(A)$, resulting in a degradation of the belief interval to a single point. Note that this is a specialization of the Dempster-Shafer belief interval.

---

[6]Probability Bounds refer to the models that deal with a probability interval limited by a lower probability and an upper probability [47].

3.1.7   Combination Rules for Belief Functions

The evidence accumulation process requires a method able to combine the bodies of independent evidence, that is, to combine beliefs from multiple observations [52].

The rule usually employed for the combination of independent bodies of evidence is called "Dempster's Rule" [45], [17]. There are, however, other rules for the combination of bodies of evidence [24], whose basic difference with respect to Dempster's rule resides in the normalization form [41], while keeping the same counter-intuitive imperfect behavior already cited. Sentz and Ferson, in [42], made a significant contribution in the compilation and analysis of the more popular rules of evidence combination, saving work for us in this book. Dempster's Rule will be described below, and in Section 4.2.3 some of the other rules will be analyzed, based on the work referred to above.

$\mathcal{B}el1 \oplus \mathcal{B}el2$[7] will represent the belief combination operation of two bodies of evidence. $\mathcal{B}el1 \oplus \mathcal{B}el2$ can be calculated from $m_1 \oplus m_2$; where $m_1 \oplus m_2$ denotes the mass functions of the combined effects of m_1 and m_2.

3.1.8   Dempster's Rule:

In the special case of a frame of discernment with just two elements, this rule was accurately defined and used by Johann Heinrich Lambert in his work *"Neues Organon"*, published in 1764; but its general formulation was only affirmed about 200 years after this, in 1967, by Dempster [45].

Dempster's Rule is composed of an orthogonal sum and a normalization[8]:

$$m_1 \oplus m_2(\mathcal{A}) = \mathcal{X} \sum_{\substack{\mathcal{B} \cap \mathcal{C} = \mathcal{A} \\ \mathcal{A} \neq \varnothing}} m_1(\mathcal{B}).m_2(\mathcal{C}) \ , \ \forall \mathcal{A} \subseteq \Theta \tag{3.12}$$

Where $\mathcal{X}$ is the normalization constant, defined as:

$$\mathcal{X} = 1/k \tag{3.13}$$

---

[7]For a question of consistency and ease of understanding, we will use "$\oplus$" to denote the combination of evidence by all the rules but ours, adding a superscript when the combination rule is different from Dempster's Rule.

[8]For those not accustomed to this notation, $m_1 \oplus m_2(\mathcal{A})$ is the same that $(m_1 \oplus m_2)(\mathcal{A})$.

And k is equal to 1 less the sum of masses after the intersections are multiplied:

$$k = 1 - \sum_{\mathcal{A}_i \cap \mathcal{B}_j = \varnothing} m_1(\mathcal{A}_i).m_2(\mathcal{B}_j) \tag{3.14}$$

Or, equivalently:

$$k = \sum_{\mathcal{A}_i \cap \mathcal{B}_j \neq \varnothing} m_1(\mathcal{A}_i).m_2(\mathcal{B}_j) \tag{3.15}$$

It can be observed that:

- The empty set occurs when we try to combine disjoint hypotheses, indicating total conflict between the hypotheses supported by the evidence.

- The commutative property of the multiplication assures the same result regardless of the order in which the functions are combined; representing the independence of the order of combination.

- The beliefs normalization step is mandatory only when the sum of all the resultant beliefs masses is smaller than 1; and this occurs when there are elements, with belief assigned to them, that are not common to all bodies of evidence.

Example 3.7 *In a multiple choice question, the set of possibilities* $\Theta = \{a, b, c, d, e\}$ *contains the correct answer, considering* $A = \{a\}$, $B = \{b\}$, $C = \{c\}$, $D = \{d\}$, *and* $E = \{e\}$. *When two people were asked what were their opinions concerning the probability of each one of the options being the correct one, the first person affirmed the following belief, resulting in first body of evidence:*

$$\mathcal{EB}_1 = (\mathcal{F}_1, m_1), \ \textit{where} \ \mathcal{F}_1 = \mathcal{A}, \mathcal{B}, \mathcal{C}, \mathcal{D}, \mathcal{E}$$

and

$$m_1(\mathcal{A}) = 0.23$$
$$m_1(\mathcal{B}) = 0.18$$
$$m_1(\mathcal{C}) = 0.28$$
$$m_1(\mathcal{D}) = 0.18$$
$$m_1(\mathcal{E}) = 0.13$$
$$m_1(\Theta) = 0$$

Note that as 100% of the belief was entirely divided among the hypotheses nothing was assigned to the "environment", Θ.

The opinion stated by the second person resulted in the second body of evidence:

$$\mathcal{EB}_2 = (\mathcal{F}_2, m_2), \text{ where } \mathcal{F}_2 = \mathcal{A}, \mathcal{B}, \mathcal{C}, \mathcal{E}, \Theta$$

and

$$m_2(\mathcal{A}) = 0.27$$
$$m_2(\mathcal{B}) = 0.17$$
$$m_2(\mathcal{C}) = 0.21$$
$$m_2(\mathcal{E}) = 0.21$$
$$m_2(\Theta) = 0.14$$

Note that as the second person preferred not to state an opinion about the possibility "$\mathcal{D}$"; and as he did not divide 100% of his beliefs among the other possibilities, the remaining was assigned to the environment, Θ.

Using Dempster's rule to make the combination:

Table 2: Combination by the Dempster's Rule

	$m_1(\Theta)$ $= 0$	$m_1(\mathcal{A})$ $= 0.23$	$m_1(\mathcal{B})$ $= 0.18$	$m_1(\mathcal{C})$ $= 0.28$	$m_1(\mathcal{D})$ $= 0.18$	$m_1(\mathcal{E})$ $= 0.13$
$m_2(\Theta) = 0.14$	0	0.0322	0.0252	0.0392	0.0252	0.0182
$m_2(\mathcal{A}) = 0.27$	0	0.0621	0	0	0	0
$m_2(\mathcal{B}) = 0.17$	0	0	0.0306	0	0	0
$m_2(\mathcal{C}) = 0.21$	0	0	0	0.0588	0	0
$m_2(\mathcal{D}) = 0$	0	0	0	0	0	0
$m_2(\mathcal{E}) = 0.21$	0	0	0	0	0	0.0273
	$\sum = 0$	$\sum = 0.0943$	$\sum = 0.0558$	$\sum = 0.098$	$\sum = 0.0252$	$\sum = 0.0455$

Resulting, before the normalization step, in the following body of combined evidence:

$$\mathcal{EB}_3 = (\mathcal{F}_3, m_3), \text{ where } \mathcal{F}_3 = \mathcal{A}, \mathcal{B}, \mathcal{C}, \mathcal{D}, \mathcal{E}$$

and

$$m_3(\mathcal{A}) = 0.0943$$
$$m_3(\mathcal{B}) = 0.0558$$
$$m_3(\mathcal{C}) = 0.098$$
$$m_3(\mathcal{D}) = 0.0252$$
$$m_3(\mathcal{E}) = 0.0455$$

And after the normalization step (obtained by dividing by the mass summation, 0.3188):

$$m_3(\mathcal{A}) = 0.30$$
$$m_3(\mathcal{B}) = 0.17$$
$$m_3(\mathcal{C}) = 0.31$$
$$m_3(\mathcal{D}) = 0.08$$
$$m_3(\mathcal{E}) = 0.14$$

From the numeric results one can perceive that hypothesis $\mathcal{C}$ is the most believed, followed by hypothesis $\mathcal{A}$. Thus, if someone desires to make a decision based upon these results, he would choose one of these options as the correct one.

The normalization constant, $\mathcal{X}$, measures the extension of what is unknown or the conflict between evidence, or even the belief assigned to Θ [45].

The value of the logarithm of the normalization constant, $log(\mathcal{X})$, is known as the "weight of conflict" (but in fact it represents both the conflict between the "specialists" and what is unknown by them) between the belief functions, being denoted by $Con(\mathcal{B}el_1, \mathcal{B}el_2)$:

$$Con(\mathcal{B}el_1, \mathcal{B}el_2) = log(\mathcal{X}) \tag{3.16}$$

The basis 10 logarithm was used to allow the mapping between $\mathcal{X} \to [1, \infty]$ and $Con(\mathcal{B}el_1, \mathcal{B}el_2) \to [0, \infty]$. This quantity deserves to be named "weight" because it can assume any value between 0 and ∞, in contrast with $\mathcal{X}$, which always assumes values equal or greater than 1 [45].

If $\mathcal{B}el_1$ and $\mathcal{B}el_2$ do not have any conflict, implying that the summation of the beliefs after the combination is 1, it follows that $Con(Bel_1, Bel_2) = 0$, since the sum of the mass of beliefs, k, is equal to 1. Similarly, if Bel_1 and Bel_2 do not have evidence in common, $Con(Bel_1, Bel_2) = \infty$, as all of the intersections will result in an empty set. If this is the case, the combination $Bel_1 \oplus Bel_2$ should not be utilized, as Dempster's Rule will produce an illogical result. Undesirable results also can be the result of the work with high values of weight of conflict.

Example 3.8 *Utilizing the data from Example 3.7, $k = 0.3188$, and consequently the normalization constant, $X = 3.1368$, and the weight of conflict, $Con(Bel_1, Bel_2) =$, will be equal to:*

$$Con(Bel_1, Bel_2) = log(X) = 0.4965$$

3.2   Relation between the Dempster-Shafer Theory and other Formal Models for Uncertain Treatment

The goal of this subsection is to indicate some of the uncertainty representation and treatment models that are contained[9] or have correspondence[10] with the Dempster-Shafer Theory. In the works cited the respective demonstrations can be found.

- Dempster-Shafer Theory → Bayesian Theory [25]:

  A belief or plausibility measure becomes a traditional probability measure, P, when all of the focal elements are singletons. In this case, $P(A) = Bel(A) = Pl(A) \forall A \in 2^\Theta$; this also results in the additive property of the probability measures.

  Any probability measure, P, arising from a finite set Θ, can be uniquely determined by a probability distribution function, $p : \Theta \to [0, 1]$ using the formula:

  $$P(A) = \sum_{x \in A} p(x) \tag{3.17}$$

  Concerning the Dempster-Shafer Theory, the function p is equivalent to the mass function, m, restricted to singletons.

---

[9]That is, they are particular cases of the Dempster-Shafer Theory.

[10]Implying that they can be converted in structures of the Dempster-Shafer Theory.

- Dempster-Shafer Theory → Possibility Theory [25]:

  When all of the focal elements are nested, that is, ordered by set inclusion, the plausibility measure becomes equivalent to the "possibility measure", and the belief measure
  equivalent to the "need measure", both arising from the Possibility Theory. The Possibility Theory is strongly connected to the Fuzzy Set Theory, since the "α-cuts" from
  this latter theory are a family of nested subsets.

- Dempster-Shafer Theory → Hints Theory [27]:

  Intuitively, a hint is an information body relative to some matter, in general imprecise,
  in the sense of not indicating a precise answer, but a range of possible answers. It
  is also frequently uncertain, in the sense that the available information allows several
  interpretations and cannot be said which of the interpretations is the correct one. The
  existence of internal conflict among hints is expected, since different interpretations can
  lead to contradictory answers. And it is also reasonable that there may be external
  contradictions among the different hints relative to the same matter.

- Dempster-Shafer Theory → Utility Theory [54]:

  The value of the result of a decision is frequently measured in monetary values, but
  people frequently exhibit preferences that are not consistent with the maximization of
  the expected monetary value. The Utility Theory accounts for this behavior, through
  the association of a personal value of the decision maker, measured in "utiles", with
  each state, s, of the utility function, $u = f(s)$, in such a way that the maximization of
  the expected utility contemplates choices consistent with this individual behavior. The
  Utility Theory satisfactorily takes the personal degree of risk acceptance into account,
  and is very useful when the preferences are not linearly related to values.

- Dempster-Shafer Theory → Value Based Systems [49]:

  The correspondence between belief functions and "Value Based Systems", "VBS", is
  that the combination of Dempster's rule corresponds to the combination operation in
  VBS, and the marginalization operation corresponds also to marginalization operation
  in VBS.

- Dempster-Shafer Theory → p-Boxes ("Probability Boxes") [21]:

There is a duality between both models [58], making it possible for one be converted to the other. But the relation is not one-to-one, since there could be several Dempster-Shafer structures corresponding to only one p-Box.

4   An Extension to the Dempster-Shafer Theory

4.1   Limitations of the Dempster-Shafer Theory

Along the years several questions have arisen regarding the Dempster-Shafer Theory (for example *vide* [51]), such as:

- What does it mean the probabilities expressed by the results?

- How do these probabilities relate to traditional probabilities?

- How can the counter-intuitive behavior of the rules of evidence combination be solved?

These questions are the consequence of two basic problems of this theory, suggested by Shafer in his original work on this theory, [45], and afterward explored by several authors, such as the ones whose articles are listed in [51]. The problems are as follows:

- Absence of an intrinsic way to represent uncertainty and conflict between the evidence in the results[1]:

  The original theory makes use of a separate measure, the "weight of conflict", to deal with uncertainty; this means that the visualization in the results does not occur naturally, making its use complicated in the sense that a relationship between this measure and the evidence must be created (for example, to disregard evidence whose weight of conflict is greater than a given value.).

---

[1]Using Dempster's rule it is possible to achieve some partial representation of ignorance in the final results, in the particular case where all the evidence to be combined exhibits previously some mass of belief attributed to the environment.

- Counter-intuitive behavior of the combination rules:

  Counter-intuitive behavior in the rules of evidence combination can be observed, when
  the evidence to be combined has a belief concentration in disjoint elements and a
  common element with low values of belief attributed to it (or even when the weight
  of conflict is high). These behaviors considerably complicate the use of the theory
  in its original form, because they imply a need to establish safeguards (such as the
  disregarding of evidence that exhibits the cited characteristics) that lead to a sub-
  optimal modeling of knowledge, avoiding, for example, some necessary changes that
  should be caused by exaggerated conflict among requisites.

As it is a trivial issue to note the absence of an intrinsic representation of the uncertainty in the results, this chapter will concentrate on showing the counter-intuitive behavior
of the combination rules. This section will use Dempster's Rule of combination for this
exemplification, and Section 4.2.3 will explore this behavior for other combination rules.

According to Smets in [50], the counter-intuitive behavior of Dempster's combination rule
was exemplified for the first time in [70][2], having been attacked by him and other authors (like
[57] and [29]) basically as a problem caused by the normalization step of the rule. We disagree
with this approach and defend that the counter-intuitive behavior is caused by the absence,
in the combination rule, of an intrinsic means for lowering the belief values proportionally to
the quantity of ignorance or conflict among the evidence, causing the attribution of a 100%
probability to a less believed, although common to the evidence, element.

Example 4.1 *Your car has broken and you called two auto-mechanics to make a diagnosis.*

*Mechanic 1 offers his opinion that there is a 99% probability of a fuel injection problem
({injection}), and 1% of probability of an electronic ignition problem ({ignition}):*

$$m_1(\{injection\}) = 0.99$$
$$m_1(\{ignition\}) = 0.01$$

Mechanic 2 assigns 99% of certainty to a command belt problem ({belt}), and 1% to an

---

[2]Although, Shafer himself had suggested it in [45].

electronic ignition problem ({ignition}):

$$m_2(\{belt\}) = 0.99$$
$$m_2(\{ignition\}) = 0.01$$

Using Dempster's Rule to combine both bodies of evidence, we have:

$$\Theta = \{injection, ignition, belt\}$$

	$m_1(\{injection\}) = 0.99$	$m_1(\{ignition\}) = 0.01$
$m_2(\{belt\}) = 0.99$ $m_2(\{ignition\}) = 0.01$	0 0	0 0.0001

Table 3: Combining by the Dempster's Rule

Resulting, after the normalization step, in:

$$m_3(\{belt\}) = 0$$
$$m_3(\{injection\}) = 0$$
$$m_3(\{ignition\}) = 1$$

$$\mathcal{B}el(\{ignition\}) = 1$$
$$\mathcal{P}l(\{ignition\}) = 1$$
$$\mathcal{I}(\{ignition\}) = [1, 1]$$

That is, there is a 100% probability of an electronic ignition problem, contradicting intuition, and causing some authors as [57] to state that it is not advisable to combine evidence with weight of conflict larger than a specified value, like 0.5 (as a rule of thumb). Other authors, such as [29] and Gerhard Paass, analyzing the Smets' article in [50], complain about this problem and also about the absence of an intrinsic way to represent the inconsistency of the evidence, and its resultant uncertainty, in the results.

4.2   The Extension to the Dempster-Shafer Theory

This extension to the Dempster-Shafer Theory consists of a new rule of evidence combination, able to solve the problems cited in Section 4.1 and a companion conceptual base for its interpretation and application.

4.2.1   Analyzing and Correcting Counter-intuitive Behavior

A key point in the process of knowledge representation is how to combine evidence that come from different sources, maintaining an adequate modeling of the uncertainty and conflict. The Dempster-Shafer Theory exhibits a counter-intuitive behavior in this matter, when the evidence to be combined possesses a high degree of conflict, or when belief is concentrated in disjoint hypotheses and there is a less-believed hypothesis in common. This counter-intuitive behavior limits the application range of the theory, and, at the same time, leads to a potential disregard of hypotheses that otherwise could add information to the system.

An analysis of the kind of phenomenon modeled in Example 4.1 is relevant. In it, we have two "specialists", and it is presumed that both possess the same credibility, thus, the fact of their divergence regarding the preferred hypothesis, in a way, discredits these hypotheses, and, at the same time, places credit in the hypothesis on which they had placed minor belief, but about which they agreed. If you keep in mind the epistemic reasoning, the final uncertainty of the result should increase, since there was disagreement regarding the hypothesis in which they placed the greatest quantity of probability, increasing the uncertainty about these hypotheses. On the other hand, if we had consulted several other specialists and they had disagreed with each other regarding the hypothesis on which they had placed most of their belief, but if all of them had agreed with the hypothesis where the least belief had been assigned, the probability of this hypothesis being the correct one must increase, while the final uncertainty of the result should decrease, *vide* Example 4.2.

Example 4.2 *Imagine a problem whose frame of discernment possesses 11 elements. We asked to 10 people to give the correct answer. Each one of these 10 people attributed most of his or her belief to a hypothesis disjoint from the others, and the least belief to the same hypothesis of the other 9 people. Based on the principle that all the people possess the same reliability, the fact of their divergence regarding the hypothesis where they placed more belief* *discredits this hypothesis, and, at the same time, increases the probability of the initially less believed hypothesis to be the right one, once all the people are "dependable" and agree about it.*

Thus, specialists agreeing about a hypothesis, give credit to it, although an attribution of 100% of the belief (as in the Dempster Rule) is exaggerated, once the divergence among more credited hypotheses and the smaller attribution of belief to a common hypothesis decreases the intrinsic value of the information.

The implementation of a new rule of evidence combination, which not only corrects this counter-intuitive effect, but also incorporates in the results the representation of uncertainty arising from conflicting hypotheses or from insufficient knowledge, can extend the Dempster-Shafer Theory, enabling the following advantages among others:

- The utilization of hypotheses where most of the beliefs are assigned to disjoint elements, without the side effect of results that contradict common sense.

- The combination of evidence that exhibits a high value of weight of conflict among them, without the counter-intuitive effects.

This new rule would expressively increase the domain of practical applications of the theory, and would promote a better utilization of the available information (as we are consulting specialists, and their opinion, even though discordant, must be considered).

This can be achieved by decreasing the value of beliefs assigned to the hypotheses, according to the degrees of ignorance and the conflict among them, and incorporating the uncertainty value arising from this ignorance or conflict in the result; as our combination rule makes possible. Part of this work has already been published in [9].

4.2.2   A New Rule of Evidence Combination

Our focus is based on the adoption of a new rule of evidence combination, which reduces beliefs in accordance with the ignorance or conflict among the evidence, attributing the belief remaining from this process to the environment and not to the common hypothesis [10]. The belief attributed to the environment in the final result constitutes a subjective uncertainty measure that emerges from the ignorance or conflict among bodies of evidence, a meta-probability that we have called "Lateo"[3] and denoted by "Λ", as an allusion to its origin, that is, the ignorance on the part of specialists, or that which was not able to be dominated and thus causes the conflict among the bodies of evidence[4]. Thus, the uncertainty arising from the unknown or conflict in the evidence is automatically incorporated into the result. Making possible:

1. The combination of evidence with most of their beliefs attributed to disjoint elements, without the side effect of counter-intuitive behavior.

2. The utilization of evidence with high conflict values, making otherwise useless evidence useful.

3. The avoidance of the need to discard evidence with a high degree of conflict, which could cause a sub-optimum modeling of the evolution of belief.

To combine two bodies of evidence, our rule is formed by the orthogonal sum, as in Dempster's Rule, divided by $(1 + log(1/k))$, that is, $(1 + Con(\mathcal{Bel}_1, \mathcal{Bel}_2))$:

$$m_1 \Psi m_2(\mathcal{A}) = \frac{\mathcal{X} \sum_{\substack{\mathcal{B} \cap \mathcal{C} = \mathcal{A} \\ \mathcal{A} \neq \varnothing}} m_1(\mathcal{B}).m_2(\mathcal{C})}{1 + log(\frac{1}{k})} \quad , \forall \mathcal{A} \subset \Theta \tag{4.1}$$

Where by $m_1 \Psi m_2(\mathcal{A})$ we are denoting the combination of evidence using our rule.

Note that $log\frac{1}{k} = log(\mathcal{X}) = Con(\mathcal{Bel}_1, \mathcal{Bel}_2)$.

The additional belief, coming from the quantity of belief reduced from the hypotheses is added to the initial environment belief, resulting in the Lateo:

$$\Lambda = m_1 \Psi m_2(\Theta) = (\mathcal{X}.m_1(\Theta).m_2(\Theta))_{|m_1(\Theta) \neq 0 \wedge m_2(\Theta) \neq 0} + 1 - \sum_{\substack{\mathcal{A} \subset \Theta \\ \mathcal{A} \neq \varnothing}} m_1 \Psi m_2(\mathcal{A}) \tag{4.2}$$

We can observe that $(\mathcal{X}.m_1(\Theta).m_2(\Theta))$ is equal to $m_1 \oplus m_2(\Theta)$ using Dempster's Rule, and our rule adds a value proportionate to the conflict between the evidence to this initial

---

[3]Latin word that means "to be hidden", "be out of sight", "be in the obscurity", "be or to be unknown".

[4]It can be observed that the "Lateo" remains coherent to which was stated by Jacob Bernoulli about probability: "Things may be contingent and uncertain or partially certain, depending on our knowledge."

belief.

Since the orthogonal sum is a quasi-associative[5] operation, if we need to work with more pieces of evidence, we should first combine them as in Dempster's Rule and then divide the result by:

$$1 + log(\frac{1}{k_1 + k_2 + ... + k_n}) \tag{4.3}$$

(where $k_1, k_2, ..., k_n$ represent the k factor of each pair of evidence combined sequentially)

After that, we can calculate the new environment belief by Equation 4.2 (see Example 4.4).

There are two cases of particular interest to exemplify the comparison between our rule and Dempster's: when combining evidence with most of their beliefs assigned to disjointed hypotheses, and when combining evidence with high weight of conflict (but without disjoint hypotheses). The first case is illustrated by the following two examples.

Example 4.3 *Using an example to compare the behavior of our rule with Dempster's for the scenario of combining evidence that has most of their beliefs assigned to disjoint hypotheses:*

Applying our rule to the same data shown in Example 4.1, we get:

$$k = 0,0001$$

$$\mathcal{X} = 10.000$$

$$log(\mathcal{X}) = 4$$

---

[5]Quasi-associativity: property of the mathematical operation that being not rigorously associative, can be divided into associative sub-operations; exhibiting, with regard to these sub-operations, behavior similar to the full associativity [42].

And thus:

$$m_3(\{belt\}) = 0$$

$$m_3(\{injection\}) = 0$$

$$m_3(\{ignition\}) = 0.2$$

$$\Lambda = m_3(\Theta) = 0.8$$

This result illustrates a more natural modeling of reasoning, since belief in the "belt" and in the "injection" continue being disregarded due to their disjunction, but the uncertainty is better represented, since the Lateo is 80%; that is, 80% of the belief becomes assigned to the environment and not to a particular hypothesis. Using these data for the plausibility function and belief interval calculation, we get:

$$\mathcal{B}el(\{ignition\}) = 0.2$$

$$\mathcal{P}l(\{ignition\}) = 1$$

$$\mathcal{I}(\{ignition\}) = [0.2 \ , \ 1]$$

Again a more realistic model of the problem is produced, since the plausibility continues to be 100%, but belief is lowered to 20%, denoting uncertainty. From an epistemic point of view, an attribution of 100% of belief to the ignition problem would not be appropriate, as happens under Dempster's Rule, simply because the mechanics disagreed on the hypotheses in which they believed the most and they agreed on a hypothesis to which they attributed little probability. It must also be noted that the larger the Lateo (that is, the ignorance or conflict between evidence), the larger or broader, also, is the belief interval, reproducing the epistemic behavior of the decrease in our capacity to define our beliefs with precision whenever our ignorance or conflict increases.

On the other hand, if new opinions coincide concerning the less believed hypothesis, this coincidence must, if we want to maintain an epistemic coherence, tend to increase the probability of the hypothesis; what we call the "confirmation effect". This behavior is also modeled in the new rule, as it can be seen in the next example.

Example 4.4 *Imagine the problem shown in the Example 4.1, but this time two more mechanics have been consulted. The first two stick to their opinions, and the other two assert* the following:

The Mechanic 3 holds for a spark plug problem ({plug}), with an 80% of belief, a battery problem with 19% ({battery}), and an ignition problem with a probability of 1%:

$$m_3(\{plug\}) = 0.80$$
$$m_3(\{battery\}) = 0.19$$
$$m_3(\{ignition\}) = 0.01$$

The Mechanic 4 bets with a 75% of probability on a fuse problem ({fuse}), 20% of probability of a lubrication problem ({lubrication}), 4% on a fuel problem ({fuel}), and 1% on an ignition problem:

$$m_4(\{fuse\}) = 0.80$$
$$m_4(\{lubrication\}) = 0.19$$
$$m_4(\{fuel\}) = 0.19$$
$$m_4(\{ignition\}) = 0.01$$

Applying the Dempster's rule, the counter-intuitive behavior remains the same:

$$m_5(\{belt\}) = 0$$
$$m_5(\{injection\}) = 0$$
$$m_5(\{plug\}) = 0$$
$$m_5(\{battery\}) = 0$$
$$m_5(\{fuse\}) = 0$$
$$m_5(\{lubrication\}) = 0$$
$$m_5(\{fuel\}) = 0$$
$$m_5(\{ignition\}) = 1$$

Applying our rule, we get as the k factor of the combination between evidence from the mechanics 1 and 2:

$$k_1 = 0.0001$$

Combining this result with the evidence from the Mechanic 3:

$$k_2 = 0.01$$

And, finally, combining this partial result with the evidence from Mechanic 4:

$$k_3 = 0.01$$

Joining all the factors:

$$k_{final} = 0.0201$$

Resulting in:

$$1 + log\frac{1}{k_{final}} = 1.70$$

And:

$$m_5(\{belt\}) = 0$$

$$m_5(\{injection\}) = 0$$

$$m_5(\{plug\}) = 0$$

$$m_5(\{battery\}) = 0$$

$$m_5(\{fuse\}) = 0$$

$$m_5(\{lubrication\}) = 0$$

$$m_5(\{fuel\}) = 0$$

$$m_5(\{ignition\}) = 0.37$$

$$\Lambda = m_5(\Theta) = 0.63$$

It is important to stress that our rule has superior performance in comparison with Dempster's rule even in the case of hypotheses with non-disjoint elements, since, whatever be the case, it will lower the beliefs attributed to the hypotheses, proportionally to the weights of conflict among them, allowing its utilization even with high values of weight of conflict, and serving to model the uncertainty and the inconsistency among specialists/opinion givers. The following example illustrates the comparison of our rule with the Dempster's one, in the case of combination of evidence with a high degree of conflict.

Example 4.5 *Using the data from examples 3.7 and 3.8, it can seen that even a relatively high value of weight of conflict $(Con(Bel_1, Bel_2) = 0.4965)$, does not make any difference when the evidence are combined using Dempster's rule, which acts in the same way that it* *would act if there were no conflict at all:*

$$m_3(\mathcal{A}) = 0.30$$
$$m_3(\mathcal{B}) = 0.17$$
$$m_3(\mathcal{C}) = 0.31$$
$$m_3(\mathcal{D}) = 0.08$$
$$m_3(\mathcal{E}) = 0.14$$

However, with the application of our rule we obtain a Lateo of $\approx 33\%$, meaning a subjective uncertainty in the results, or an assignment of $\approx 33\%$ of the total belief to the environment, accompanied by the reduction of the beliefs of the other elements, denoting the uncertainty derived from the conflicting evidence:

$$m_3(\mathcal{A}) = 0.200$$
$$m_3(\mathcal{B}) = 0.114$$
$$m_3(\mathcal{C}) = 0.207$$
$$m_3(\mathcal{D}) = 0.053$$
$$m_3(\mathcal{E}) = 0.094$$
$$\Lambda = m_3(\Theta) = 0.332$$

It can be noticed that the relative position of the elements remains intact, but the beliefs are reduced proportionally to the weight of conflict, as usually happens when we intuitively process conflicting evidence.

On the other hand, the utilization of the Lateo makes it easy to discern if the combined evidence are sufficient for a rational decision be made, as we show in the Example 4.6.

Example 4.6 *Let's consider a frame of discernment composed of 5 elements whose combination of evidence resulted in these values:*

$$m_4^\Lambda(\mathcal{A}) = 0.233$$
$$m_4^\Lambda(\mathcal{B}) = 0.149$$
$$m_4^\Lambda(\mathcal{C}) = 0.290$$
$$m_4^\Lambda(\mathcal{D}) = 0.030$$
$$m_4^\Lambda(\mathcal{E}) = 0$$
$$\Lambda = m_4^\Lambda(\Theta) = 0.298$$

The value of the Lateo in this example, would not allow a rational choice be made, because if it were attributed to any one of the hypotheses, it would make it the most credible. Nevertheless, the values attributed to the hypotheses suggest what can be deduced, given the information from the evidence used. Thus, these bodies of evidence indicate that the beliefs attributed by the sources to each hypothesis is insufficient, in this case, for a rational decision be made.

4.2.2.1   Modeling Disruptive Evidence:

Computational systems using the Theory of Evidence under Dempster's Rule exhibit either counter-intuitive behavior, or discard input with weights of conflict above a particular level. The discard of the inputs carries to a sub-optimum modeling of the beliefs evolution, since the discarded inputs can, for example, represent an inflection in the belief evolution, and all the posterior entrances with similar or bigger weights of conflict will also be refused, resulting in total ignorance in the change of direction of the tendency of the belief evolution. This input discard leads to a sub-optimum modeling of the evolution of belief, since the discarded inputs may, for example, represent a radical change in the evolution of belief. All further inputs with similar or larger weights of conflict will also be refused, resulting in total ignorance of the belief evolution. It can be imagined that in certain areas of knowledge where radical changes in the paradigm occur with some frequency, such as in tropical medicine or orthopedics, this behavior represents a problem. Our rule corrects this problem, accepting the inputs and adding to the result, through the Lateo, the uncertainty representation caused by them. This characterizes the radical change of the paradigm; and as the input that agree with the new paradigm increase, the Lateo, or in other words, the uncertainty attributed to the environment, will decrease, characterizing the new paradigm sedimentation as the most acceptable solution and thus reproducing the epistemic reasoning.

4.2.3 Validation of the Dempster-Shafer Theory Extension

In this section the resultant extension will be validated, with its behavior examined in relation to several mathematical properties and compared with other rules of evidence combination. Special care has been employed to explain the conceptual basis of its behavior.

Given the nonexistence of a well-founded mathematically established epistemology of uncertainty, a formal demonstration of a given rule of combination or extension to a formal model adequate to deal with beliefs becomes impossible.

Using simplistic explanations, the scientific community has been able to justify the adoption of new rules of combination, or extensions to the Theory of Evidence, using examples and counter-examples, demonstrating the achievement of counter-intuitive results when other rules are applied, and the adequacy of the new approach. In this way, the new rule becomes valid until the appearance of new counter-examples that demonstrate new situations where the counter-intuitive results are achieved. Thus this process culminates in the creation of new rules or extensions, beginning the cycle again.

From a purely mathematical point of view, one could consider arbitrary operations for the combination of evidence. The combination rules could occupy a range of mathematical operations ranging between total conjunction and total disjunction, varying by different degrees between these two extremes[6] [20]. However, as it is desirable that this combination occur in a sensitive and significant way, there are some requirements tied to epistemic analysis and to common sense that the aggregation operation must satisfy.

There are several analysis and validation criteria for the rules of evidence combination; one of these verifies if the rule complies with the criterion of rational bets already mentioned; another, recently more in use owing to its better developed mathematical basis (examples, [42] and [21]), consists of an analysis of the total or partial compliance of several mathematical properties, along with a conceptual foundation for this behavior and a comparison with

---

[6]These operation classes are named by Dubois and Prade in [20] as "conjunctive pooling", "disjunctive pooling", and "trade off", respectively.

several contemporary combination rules. Support for all of the mathematical properties is
not absolutely necessary, since there are pro and con arguments for most of them. What
seems to be reasonable in one situation can not always be entirely reasonable in another
context, or when we change the kind of information, *vide* [21]. Nevertheless, it is desirable
to consider under what conditions these aggregation operations satisfy the mathematical
properties [21].

The methodology followed in this work is similar to the one described in Ferson *et al* [21]
and used in the article of Sentz and Ferson [42][7]. A small difference in relation to this work
consists in the non-utilization of all mathematical properties used in [42], given that there was
no need to include some of the properties that deal, specifically, with the treatment of some
kinds of uncertain numbers (as the work with these uncertain numbers, like the "probability
intervals" or the "p-Box", is outside the scope of this work). Also, in this present book, we
have not explored in detail the several rules of combination of evidence available, since Sentz
and Ferson in [42] and Ferson *et al.* in [21] have already researched, compiled and analyzed
the more important aggregation rules. So we will summarize these rules and cite behavior
studied by them when it is necessary.

We will now proceed to show that the proposed new rule of combination of evidence
constitutes a significant advance over what can be regarded as an extension to the Demspter-
Shafer's Theory, since it solves two classical problems of the theory: the counter-intuitive
behavior of the combination rule and the lack of the representation of ignorance and conflict
in the result.

4.2.3.1   Comparison with some other Rules of Evidence Combination:

Now we will introduce some of the most important rules of combination of evidence, and
then we will compare their behavior with the proposed new rule.

- **Discount and Combine Rule:**

  This is a "trade off" kind of method proposed by Shafer in his work [45]. It is based
  on the principle that when there is conflict among the bodies of evidence a referee can

---

[7]The apparent contradiction between the dates of the description of the methodology – 2003 in [21] – and
its use - 2002 in [42] – is not significant, as Ferson *et al* had access to the methodology and used it in [21]
before its publication).

first discount the evidence's beliefs and then combine them using a discount function. This discount function must consider the sources of absolute confidence, which implies that the referee is able to distinguish the degrees of confidence of the specialists, sensors, or from other information input, as well as being able to express this distinction mathematically. Afterward, similar methods with denominations like Mixture or Average Rule [21] were developed. The basic difference among the newer "discount and combine" methods and the idea proposed by Shafer is that Shafer proposed to operate on belief functions, $\mathcal{B}el$, while the new methods work with the belief masses, m. Due to their similarities regarding the observations that we will make, we will only deal with operations treating of belief masses in this section.

Mathematically, these methods consist in discount the beliefs by a factor $(1 - \alpha_i)$ (vide Equation 4.4), denominated by Shafer as a "confidence degree", where "i" indicates the factor associated to a belief in particular, and $0 \leq \alpha_i \leq 1$. After the discount, the final result is achieved by averaging the belief functions. This average can be the one shown in Equation 4.5, or other kinds of averages, such as the weighted one, can be employed.

$$m^{\alpha_i}(A) = (1 - \alpha_i)m(A) \tag{4.4}$$

$$\overline{m(A)} = \frac{1}{n}(m^{\alpha_1}(A) + m^{\alpha_2}(A) + \ldots + m^{\alpha_n}(A)) \tag{4.5}$$

Where $m^{\alpha_i}(A)$ represents the discounted belief mass, and $\overline{m(A)}$ the belief mass "result", obtained through an average of the "n" elements corresponding to the subsets in the frame of discernment.

Thus, this kind of approach, suggested initially by Shafer, treats the counter-intuitive behavior described in Section 4.1 as if the cases with high degrees of conflict could be solved through the average of the assigned beliefs. Shafer also advocates (see [45]), along with this solution, the possibility of discarding evidence with high degree of conflict, "if this would be reasonable in a particular situation". In our opinion both solutions move away from the epistemic reasoning:

 - If a source of evidence has been consulted from an epistemic point of view, this means that it has some relevance; if this is not the case, it should not have been consulted in the first place.

   Why would a source be consulted if there was no intention of using its information?

Thus, the fact that beliefs arising from a given source disagree, even to a high degree, rather than being modeled as a contempt for the information, should be modeled as an increase in the degree of uncertainty or ignorance in the final result, which is provided for in our rule, through an automatic increase of the Lateo.

– On the other hand, trying to solve the conflict through averaging the beliefs, even using some kind of weighting, also leads to non-epistemic results, since, again, the conflict must be modeled as an increase in the degree of uncertainty or ignorance in the results (which the Lateo provides for in our rule), and not as a need to arrive at a consensus using the average. Example 4.7 illustrates this point.

Example 4.7 *Imagine again Example 4.1 (where your car broke and you called two independent mechanics to diagnose the problem). But now, both mechanics have assigned 100% of their beliefs to disjoint hypotheses, with Mechanic 1 assigning 100% of probability to an electronic injection problem ({injeção}), and Mechanic 2 betting with 100% of certainty that the problem is the belt of the valve command ({belt}):*

$$m_1(\{injection\}) = 1$$
$$m_2(\{belt\}) = 1$$

Combining both bodies of evidence by the Dempster's Rule, one get:

$$\Theta = \{injection, belt\}$$

	$m_1(\{injection\}) = 1$	$m_1(\{belt\}) = 0$
$m_2(\{injection\}) = 0$	0	0
$m_2(\{belt\}) = 1$	0	0

Table 4: Combining by the Dempster's Rule

Resulting after the normalization step in:

$$m_3(\{belt\}) = 0$$
$$\mathcal{B}el(\{belt\}) = 0$$
$$\mathcal{P}l(\{belt\}) = 0$$
$$\mathcal{I}(\{belt\}) = [0, 0]$$

$$m_3(\{injection\}) = 0$$
$$\mathcal{B}el(\{injection\}) = 0$$
$$\mathcal{P}l(\{injection\}) = 0$$
$$\mathcal{I}(\{injection\}) = [0, 0]$$

Applying Dempster's Rule directly would result in the improbability of either of the results being the right one, that is, both pieces of evidence must be wrong. However, through an epistemic analysis we can see how counter-intuitive this result is, given the fact that when credible sources disagree (if they were not worthy of credit, they should not be consulted in the first place) it does not mean that both are wrong, but that there could be a whole range of explanations, for instance:

— The background or the data available led one of them to reason incorrectly and the other one reasoned correctly.

— The incomplete information on which they based their reasoning did not suggest other alternatives, so the hypothesis indicated by each one was the only explanation they could come up with (the fact of being "specialists" or people "worthy of credit" does not mean that they cannot be mistaken or fail) among the solutions that could be correct.

Conversely, the application of a procedure using averages upon the beliefs would not take us toward an epistemic result, because the final result would fail to indicate the uncertainty or ignorance, arising from the high degree of conflict in the evidence (vide Example 4.8). Using epistemic reasoning, when one receives conflicting evidence, one decreases both the credibility of the final result (taken care of in the proposed rule by an increase in the Lateo) and belief in the separate hypotheses (which is taken care of in the proposed rule by reducing the beliefs proportionally to the weight of conflict).

Example 4.8 *Considering the equitable distribution of the "degree of confidence" (in Shafer's sense) between the results, one would have:*

$$m_3(\{belt\}) = 0.5$$
$$\mathcal{B}el(\{belt\}) = 0.5$$
$$\mathcal{P}l(\{belt\}) = 0.5$$
$$\mathcal{I}(\{belt\}) = [0.5\ ,\ 0.5]$$

$$m_3(\{injection\}) = 0.5$$
$$\mathcal{B}el(\{injection\}) = 0.5$$
$$\mathcal{P}l(\{injection\}) = 0.5$$
$$\mathcal{I}(\{injection\}) = [0.5\ ,\ 0.5]$$

Observe that the rules of "discount and combine" and "mixture" or "average" make use of a "forced consensus"; eliminating the representation of the ignorance or the conflict between the specialists from the final result.

- **Yager's Rule:**

  The Yager's Rule represents, according to [42], the most prominent form of the aggregation of evidence (we believe, however, that our new rule, proposed in this work, possesses even better behavior), as it is, according to them, honest in an epistemic sense, without changing the evidence to achieve a normalization that expurgates the conflict. Above, it has been argued and demonstrated in our rule that the problem does not reside in the normalization operation itself, but in the form in which it it has been applied up to the time of our work. This rule, however, is conceptually similar to the one created by us, making it worthwhile to explain it here.

  Disregarding new denominations[8] the idea of Yager is that counter-intuitive behavior

---

[8]When authors create new rules of combination, they generally also create new denominations for the artifacts previously named by Dempster and Shafer, but with little justification in what regards for the difference in concepts. To clarify understanding and attempt uniformity, in our work we have used the same denominations and notations used in the Dempster-Shafer Theory when referring to artifacts that possesses conceptual identity with the ones named by this theory.

is caused by the normalization operation, and that by eliminating normalization this would be solved (remembering that we have already stated that we do not agree with this reasoning). An interesting point of Yager's proposal, however, is that he glimpsed the need to provide some kind of representation of ignorance and conflict in the result, or what he called "ignorance degree". Yager formulated his idea in the following form:

$$m_1 \oplus^Y m_2(\mathcal{A}) = \sum_{\mathcal{B} \cap \mathcal{C} = \mathcal{A}} m_1(\mathcal{B}).m_2(\mathcal{C}) \qquad (4.6)$$

Where by "$m_1 \oplus^Y m_2$" one denotes the combination of the mass of beliefs using Yager's Rule.

Yager states that Dempster's Rule changes, by the normalization operation, the evidence by assigning the conflicting belief mass to the empty hypothesis – we consider this conceptually debatable, as it is our understanding that the normalization operation divides the conflicting mass between hypotheses that remain with belief in the combination using Dempster's Rule. Thus Yager allows for the possibility that in his formula a mass of belief greater than zero may be attributed to the empty hypothesis (Equation 4.7).

$$m_1 \oplus^Y m_2(\varnothing) \geq 0 \qquad (4.7)$$

The value of $m_1 \oplus^Y m_2(\varnothing)$ can be calculated by adding the product of the masses of belief that do not intersect (which actually represents the mass of conflict or ignorance of the evidence), vide Equation 4.8, maintaining uniformity with Equation 4.7.

$$m_1 \oplus^Y m_2(\varnothing) = \sum_{\mathcal{B} \cap \mathcal{C} = \varnothing} m_1(\mathcal{B}).m_2(\mathcal{C}) \qquad (4.8)$$

The value of the ignorance or conflict between the bodies of evidence, denominated by Yager as "ignorance degree", $m^Y(\Theta)$, is represented in the final result as the sum of the mass of belief attributed by Yager to the empty hypothesis with the belief mass attributed to the environment, or frame of discernment, as is exhibited in Equation 4.9.

$$m^Y(\Theta) = m_1 \oplus^Y m_2(\Theta) + m_1 \oplus^Y m_2(\varnothing) \qquad (4.9)$$

An example of the behavior of Yager's Rule is given in Example 4.9, using data from Example 4.1. This same data will also be used to show the behavior of our rule (vide

Example 4.3).

Example 4.9 *Using the data from the evidence from Example 4.1 and combining them using Yager's Rule, one would obtain:*

$$m_3^Y(\{belt\}) = 0$$
$$m_3^Y(\{injection\}) = 0$$
$$m_3^Y(\{ignition\}) = 0.0001$$
$$m_3^Y(\Theta) = 0.9999$$

$$\mathcal{B}el(\{ignition\}) = 0.0001$$
$$\mathcal{P}l(\{ignition\}) = 1$$
$$\mathcal{I}(\{ignition\}) = [0.0001 \ , \ 1]$$

As it can be deduced from these results, the combination using the Yager's Rule, although providing for a representation of the ignorance and conflict of the evidence in the result, exhibits a series of counter-epistemic behaviors:

- *Reduction by several orders of the belief attributed to the ignition, even considering the fact that all computed evidence agreed regarding the belief assigned to it. We believe that to be consistent in an epistemic sense, the belief in ignition would have to be reinforced (that is, increased to more than 1% and not reduced to 0.01% as it was).*

- *If other evidence had been computed, even though they agreed with the same numeric value regarding the belief attributed to the {ignition}, the final belief in {ignition} would still be smaller (that is, the concordance instead of increasing the belief, would decrease it).*

Applying the proposed rule to this same data (Example 4.3), none of these counter-intuitive behaviors are observed.

- **Inagaki's Rule:**

The basic idea behind Inagaki's Rule is the definition of a continuous and parametrical class of evidence combination operations, encompassing both Yager's and Dempster's rules.

Inagaki considers that any rule of evidence combination can be expressed in the form of Equation 4.10, taking into account equations 4.11 and 4.12. This is not true, however, because our rule can not be expressed using these equations (vide Example 4.10).

$$m_1 \oplus^I m_2(\mathcal{A}) = m_1 \oplus^Y m_2(\mathcal{A}) + f(\mathcal{A}).m_1 \oplus^Y m_2(\varnothing), \forall \mathcal{A} \neq \varnothing \tag{4.10}$$

$$\sum_{\substack{\mathcal{A} \subseteq \Theta \\ \mathcal{A} \neq \varnothing}} f(\mathcal{A}) = 1 \tag{4.11}$$

$$f(\mathcal{A}) \geq 0 \tag{4.12}$$

Thus, the function "f" can be considered as a scale factor for $m_1 \oplus^Y m_2(\varnothing)$; being the conflict, "$k_{UI}$", defined in Equation 4.13.

$$k_{UI} = \frac{f(\mathcal{A})}{m_1 \oplus^Y m_2(\mathcal{A})} \tag{4.13}$$

Inagaki restricts the scope of his combination rules class to the rules that satisfy Equation 4.14, for all elements $\mathcal{A}, \mathcal{B}$ different from the empty hypothesis and of the frame of discernment.

$$\frac{m_1 \oplus^I m_2(\mathcal{A})}{m_1 \oplus^I m_2(\mathcal{B})} = \frac{m_1 \oplus^Y m_2(\mathcal{A})}{m_1 \oplus^Y m_2(\mathcal{B})} \tag{4.14}$$

From the equations 4.10, 4.11, and 4.13, Inagaki extracts what he calls the "Unified Combination Rule", which will be denoted for "$m_1 \oplus^{UI} m_2(\mathcal{A})$", as it can be seen in the equations 4.15, 4.16, and 4.17.

$$m_1 \oplus^{UI} m_2(\mathcal{A}) = [1 + k_{UI}.m_1 \oplus^Y m_2(\varnothing)].m_1 \oplus^Y m_2(\mathcal{A}), \forall \mathcal{A} \neq \varnothing \tag{4.15}$$

$$m_1 \oplus^{UI} m_2(\Theta) = [1 + k_{UI}.m_1 \oplus^Y m_2(\varnothing)].m_1 \oplus^Y m_2(\Theta)$$
$$+[1 + k_{UI}.m_1 \oplus^Y m_2(\varnothing) - k_{UI}].m_1 \oplus^Y m_2(\varnothing) \tag{4.16}$$

$$0 \leq k_{UI} \leq \frac{1}{1 - m_1 \oplus^Y m_2(\varnothing) - m_1 \oplus^Y m_2(\Theta)} \tag{4.17}$$

The "k_{UI}" parameter is used to define the normalization degree. However, both the works of Inagaki and the literature consulted for the present work fail to establish a conceptually well-founded procedure for determining k_{UI}. Tanaka and Klir in [55], suggest determining k_{UI} using experimental data, simulations, or by consulting the expectations of a specialist. In our opinion, this assertion of Tanaka and Klir is debatable, mainly from a conceptual point of view, since it implies the admission of a possibility of a rule that encompass the behaviors of all the rules, from Yager's to Dempster's (or of some other more extreme rule, regarding the equitable division of the conflict and ignorance between beliefs, achieved with other values of k_{UI}). As we have already explained, we found both rules debatable in relation to various aspects of their behavior. From a pragmatic point of view, given the fact that experimentation is not always possible, this procedure would introduce the questionable aspect of the election of a specialist whose expectation would be used as a standard; or further, we would find k_{UI} from the results of experiments or simulations whose analysis would once again depend on an arbitrator. To us, there is no need to vary the influence of the degrees of ignorance or conflict between the evidence upon the resultant beliefs assigned to the focal elements, even if this leads to the counter-intuitive behaviors illustrated in the examples given earlier. What it is actually important is to represent uncertainty or the degree of ignorance in the final result, so that, according to the particular situation, decisions about the credibility, confidence or the utility of the beliefs can be attributed to these elements.

It is important to stress that Inagaki's Rule is not able to model the proposed rule in a generic form (but only in particular cases). In order to make it generic, one must change the k_{UI} value for each new combination of evidence which is not supported by his argument. Example 4.10 illustrates this behavior.

Example 4.10 *Imagine the combination of 3 bodies of evidence that possess the following masses of belief:*

$$
\begin{array}{lll}
m_1(\mathcal{A}) = 0.23 & m_2(\mathcal{A}) = 0.27 & m_3(\mathcal{A}) = 0.27 \\
m_1(\mathcal{B}) = 0.18 & m_2(\mathcal{B}) = 0.17 & m_3(\mathcal{B}) = 0.17 \\
m_1(\mathcal{C}) = 0.28 & m_2(\mathcal{C}) = 0.21 & m_3(\mathcal{C}) = 0.21 \\
m_1(\mathcal{D}) = 0.18 & m_2(\mathcal{D}) = 0 & m_3(\mathcal{D}) = 0 \\
m_1(\mathcal{E}) = 0 & m_2(\mathcal{E}) = 0.21 & m_3(\mathcal{E}) = 0.21 \\
m_1(\Theta) = 0.13 & m_2(\Theta) = 0.14 & m_3(\Theta) = 0.14
\end{array}
$$

Combining the first 2, using Dempster's Rule, $m_4^D = m_1 \oplus m_2$, we get:

$$m_4^D(\mathcal{A}) = 0.324$$
$$m_4^D(\mathcal{B}) = 0.191$$
$$m_4^D(\mathcal{C}) = 0.336$$
$$m_4^D(\mathcal{D}) = 0.087$$
$$m_4^D(\mathcal{E}) = 0$$
$$m_4^D(\Theta) = 0.062$$

Combining them by the proposed rule, $m_4^\Lambda = m_1 \Psi m_2$, would result in:

$$m_4^\Lambda(\mathcal{A}) = 0.211$$
$$m_4^\Lambda(\mathcal{B}) = 0.125$$
$$m_4^\Lambda(\mathcal{C}) = 0.219$$
$$m_4^\Lambda(\mathcal{D}) = 0.056$$
$$m_4^\Lambda(\mathcal{E}) = 0$$
$$\Lambda = m_4^\Lambda(\Theta) = 0.389$$

To obtain the same result as the proposed rule, when using Inagaki's Rule, one would have to choose $k_{UI} = 1.742$.

Doing the combination of m_1 and m_2 com m_3, by the Dempster's Rule, $m_5^D = m_1 \oplus m_2 \oplus m_3$:

$$m_4^D(\mathcal{A}) = 0.401$$
$$m_4^D(\mathcal{B}) = 0.180$$
$$m_4^D(\mathcal{C}) = 0.356$$
$$m_4^D(\mathcal{D}) = 0.037$$
$$m_4^D(\mathcal{E}) = 0$$
$$m_4^D(\Theta) = 0.026$$

And using the proposed rule, $m_5^\Lambda = m_1 \Psi m_2 \Psi m_3$:

$$m_4^\Lambda(\mathcal{A}) = 0.333$$
$$m_4^\Lambda(\mathcal{B}) = 0.149$$
$$m_4^\Lambda(\mathcal{C}) = 0.295$$
$$m_4^\Lambda(\mathcal{D}) = 0.030$$
$$m_4^\Lambda(\mathcal{E}) = 0$$
$$\Lambda = m_4^\Lambda(\Theta) = 0.193$$

These values would have to be at $k_{UI} = 2.250$, to be obtained by Inagaki's Rule. So we can see that Inagaki's Rule is not able to model, generically, the behavior of the proposed rule, since it is not able to lower the beliefs automatically in proportion to the degree of conflict of each pair of bodies of evidence to be combined. Thus, to reproduce a behavior similar to our rule, it would be necessary, in a generic case, to adopt a new value for k_{UI}, for each new evidence to be combined, which implies, using Inagaki's method, the adoption of a new aggregation rule whenever a new combination of evidence combination is necessary.

The rules of evidence combination hitherto explained have their main line of reasoning in common to with the proposed rule. Now let us consider some other rules that are a little further away from the line of argument we have proposed.

- **Zhang's Central Combination Rule:**

Zhang's basic Idea is to discount the evidence by a measure of the intersection between the focal elements, defined as being the reason between cardinality of the intersection between two elements and product of the cardinality of these elements. In the Equation 4.18 this measure is denoted by $r(\mathcal{A}, \mathcal{B})$, and for $|\mathcal{P}|$ we are denoting the cardinality of $\mathcal{P}$.

$$r(\mathcal{A}, \mathcal{B}) = \frac{|\mathcal{A} \cap \mathcal{B}|}{|\mathcal{A}|.|\mathcal{B}|} \tag{4.18}$$

The belief mass of a certain element would then be expressed by the Equation 4.19, where k represents a constant of re-normalization that makes the sum of all the masses equal to 1.

$$m(\mathcal{A}) = k \sum_{\mathcal{B} \cap \mathcal{C} = \mathcal{A}} \frac{|\mathcal{B} \cap \mathcal{C}|}{|\mathcal{B}|.|\mathcal{C}|} m_1(\mathcal{B}).m_2(\mathcal{C}) \tag{4.19}$$

Our main objection to this rule is that every time that the cardinality of all of the focal elements is 1 (as in Example 4.1), or when $|\mathcal{B} \cap \mathcal{C}| = |\mathcal{B}|.|\mathcal{C}|$, the Zhang's Rule exhibits the same behavior as Dempster's Rule, with all the deficiencies associated with it.

- **Dubois and Prade's Consensual Disjunctive Rule:**

This rule defends a means of analysis that is very diverse from the previous ones, where instead of ending up with a result representing the combination of the evidence, one ends up with a group of belief unions as the final result, and draws the conclusion from this group. Vide Equation 4.20.

$$m_1 \cup m_2(\mathcal{A}) = \sum_{\mathcal{B} \cup \mathcal{C} = \mathcal{A}} m_1(\mathcal{B}).m_2(\mathcal{C}) \ \ , \ \forall \ \mathcal{B} \in 2^{\Theta} \tag{4.20}$$

Note that this kind of rule supplies a group of belief unions as a result. Both the interpretation of this group of belief unions, as well as the beliefs values themselves, are questionable, as we show in Example 4.11.

Example 4.11 *Using the data from Example 4.1, and applying the Consensual Disjunctive Rule of Dubois and Prade, one would arrive at:*

$m_1 \cup m_2(\{belt\} \cup \{belt\}) = 0$

That is, the belief that the problem may be the belt is 0.

$m_1 \cup m_2(\{injection\} \cup \{injection\}) = 0$

That is, the belief that the problem may be the injection is 0.

$m_1 \cup m_2(\{belt\} \cup \{injection\}) = 0.9801$

That is, the belief that the problem may be the belt or the injection is 0.9801.

$m_1 \cup m_2(\{belt\} \cup \{ignition\}) = 0.0099$

That is, the belief that the problem may be the belt or the ignition is 0.0099.

$m_1 \cup m_2(\{injection\} \cup \{ignition\}) = 0.0099$

That is, the belief that the problem may be the injection or the ignition is 0.0099.

$m_1 \cup m_2(\{ignition\} \cup \{ignition\}) = 0.0001$

That is, the belief that the problem may be the ignition is 0.0001.

$m_1 \cup m_2(\{belt\} \cup \{injection\} \cup \{ignition\}) = 1$

That is, the belief that the problem may be the belt or the injection or the ignition is 100%.

The biggest problem of this rule is how to interpret its results. The literature lacks a methodology that provides for this. In its absence, one can establish a series of questions about the group of belief unions achieved, both regarding its conceptual interpretation and in what concerns the beliefs values themselves, for example:

As was argued in Section 4.2.1, when we questioned the Discount and Combine Rule, and in Example 4.7, the lack of agreement among the more trusted hypotheses serves to discredit them. Using the Disjunctive Consensus Rule, we obtain several results questionable in an epistemic sense, such as:

- *If the belief ({belt} $\cup$ {belt}) = 0 and the ({injection} $\cup$ {injection}) = 0 also, we do not see epistemic justification that the belief ({belt} $\cup$ {injection}) can be equal to 0.9801.*

- *The same applies to the beliefs ({belt} $\cup$ {ignition}) = 0.0099 e ({injection} $\cup$ {ignition}) = 0.0099, we to not see an epistemic excuse for the union of beliefs that when combined among themselves result in 0, but joined to a third hypothesis, whose union with itself be different from 0, can result in a an increase in the belief growth of this third hypothesis.*

  How can this behavior be justified by the combination of the knowledge of the specialists?

  As we see it cannot.

- *How can two hypotheses have a belief different from zero when united, if the beliefs within each are zero?*

  Again, we can not manage to obtain a well-based epistemic answer.

- **Other Rules of Evidence Combination:**

  Although there are several others rules of combination of evidence, these rules have not been discussed here because their scope are is not within the purview of the present work. Examples:

  - Yao and Wong's Rule: because it only deals with the qualitative analysis of belief.

  - Smet's Rule: since it is Dempster's Rule with a new conceptual framework and has the same deficiencies, cited in 4.1, as the original rule.

  - The Envelope, Imposition and similar rules: because they work specifically with the "p-Boxes".

4.2.3.2 The behavior of rules in the light of the mathematical properties relevant to the combination of evidence:

Several mathematical properties that have an impact on the epistemic behavior of the combination of evidence will now be explained. The need for total, partial or non-assistance to the several properties when the rules are applied to various types of evidence combination, relevant to each one of these properties, will be explained.

1. **Idempotence:**

   An aggregation rule obeys idempotence if the combination of two bodies of evidence with the same belief value in all elements has, as a result, the same belief values as these elements. Vide Equation 4.21.

   $$m_1 \oplus m_1 = m_1 \tag{4.21}$$

   In what situations can we expect to use idempotence?

   Keeping in mind the premise of the independence of the sources of evidence in the sense that a source does not influence the belief masses of the other sources, idempotence cannot be used generically, but only in specific situations. Let's identify and analyze all the possible kinds of evidence combinations:

   - Combination of two vacuous evidence:

     The vacuous function is a function that represents either total ignorance or lack of knowledge. It is represented for an evidence in which $m(\Theta) = 1$. Axiomatically, it is not possible to obtain more than a 100% of ignorance, and as the combination of two bodies of evidence with 100% of ignorance, in an epistemic sense, does not lead to the decrease of this ignorance, the combination of two vacuous bodies of evidence must also result in vacuous evidence. That is, the combination of vacuous evidence must obey the idempotence property.

   - Combination of evidence with all the belief concentrated in just one hypothesis:

     In an analogous way, by axiom, it is not possible to obtain more than 100% of probability, and because common sense shows that the combination of two pieces of evidence that are certain with respect to the same hypothesis should not decrease this certainty, this combination must also obey the idempotence property.

- Combination between evidence with neither ignorance and nor belief concentrated in only one of the hypotheses, but with different beliefs assigned to the hypotheses:

  Common sense (epistemic way of thinking) tells us that there is a "confirmation effect" when one receives evidence that agrees, regarding the beliefs attributed to the hypotheses, from two diverse sources. This confirmation effect has the outcome of making the agreement among beliefs reinforce the most believed hypothesis, as Example 4.12 makes clear. Therefore, this type of combination cannot obey to idempotence.

Example 4.12 *Imagine that your car broke down and you consulted two independent auto-mechanics. The both said that there was an 80% probability that the defect was the ignition and 20% that it was the battery. Using epistemic sense, the coincidence of opinions should reinforce the belief that the ignition was the problem and, as the sum of all beliefs must result in 100%, the belief in the battery has to decrease.*

- Combination of evidence that do not possess ignorance and without the belief concentrated in just one hypothesis, but with the belief divided equally between the hypotheses:

  In a form similar to the previous case, the concordance of the evidence would tend to confirm the beliefs in each hypothesis, but as the sum of the beliefs must result in 100%, by axiom, and the original beliefs are equally divided between the hypotheses, the confirmation effect does not have a means to increase the specific beliefs of either hypothesis, leaving the beliefs of the combination equal to the previous ones. That is, this kind of combination must obey idempotence, vide Example 4.13.

Example 4.13 *Imagine that your car broke down and you consulted two independent auto-mechanics; now, however they attributed a 50% probability to an ignition defect and a 50% probability to a battery problem. Using epistemic reasoning, the coincidence of the opinions with the beliefs distributed equally between the hypotheses should not be able to cause a trend toward any of them in particular, neither contribute to the knowledge growth (unless by the fact of confirming that there are several equally probable options, or that the available information is insufficient).*

- Combination between evidence that possess ignorance, but that also has belief attributed to one hypothesis, or to more than one but equally distributed among them:

  According to the same confirmation principle, the coincidence must decrease the uncertainty in the final result (the Lateo, in the case of our rule) and increase beliefs in all focal elements equally. Thus, it will not obey idempotence, except in the case of rules that lack a means of representing uncertainty in the combination result (we consider this lack as a failure of these rules).

- Combination between evidence that has both ignorance and belief attributed to some of their hypotheses in an uneven way:

  Continuing on with the confirmation principle, uncertainty in the final result must decrease and the attributed beliefs must grow proportionally to their values. That is, the closer the agreement among the bodies of evidence, the more it should increase the beliefs in the final result, not conforming to idempotence.

It is worth stressing that our rule models all of these behaviors; Yager's Rule does not; Dempster's Rule does, but without considering representation or effect of the uncertainty in the result.

2. **Commutativity:**

   Commutativity is characteristic of aggregation rules where the order of combination of evidence does not change the result. With commutativity, the result of the combination is independent of the combination order, as is stated in Equation 4.22.

   $$m_1 \oplus m_2 = m_2 \oplus m_1 \qquad (4.22)$$

   This generic assistance to the commutativity makes sense. We have not observed any case that justifies the non-observance of commutativity. Our rule, as well as Yager's and Dempster's, are all commutative. Bayes Conditioning Rule is not.

3. **Continuity:**

   If we have two different evidence, m_1 and m_2, and a third evidence, m_{1B}, which is almost equal to m_1, which we have denoted by $m_1 \approx m_{1B}$, it is reasonable to come up with what is expressed by Equation 4.23.

$$m_1 \oplus m_2 \approx m_{1B} \oplus m_2 \tag{4.23}$$

In other words, it is reasonable to require that a small change in one of the bodies of evidence will result in just a small change in their combination, when we work with belief punctual values. Ferson *et al*, in [21], argue that in the case of hypotheses with probability intervals (in their paper they discuss the use of the Theory of Dempster-Shafer working with uncertain numbers in general), continuity becomes unnatural when an intersection occurs between probability intervals and when the intersection is gradually decreased until it disappears, at this point, the continuity will not make sense anymore.

The proposed rule also satisfies these continuity behaviors.

4. **Associativity:**

   To say that an aggregation rule has the associativity property implies that the result of the evidence combination does not depend on the order in which this combination is processed, as stated by the Equation 4.24.

$$(m_1 \oplus m_2) \oplus m_3 = m_1 \oplus (m_2 \oplus m_3) \tag{4.24}$$

   It is natural to expect that the result of the evidence aggregation does not depend on the order with which we aggregate them. But the assistance to this property is not essential anymore, provided that the rule be, at least, "quasi-associative" [42]. This quasi-associativity implies that we might be able to model the associative behavior by dismembering a rule which is not associative in associative sub-operations which produce an associative behavior in the result [42]. Example of this is the rule that is proposed in this work, whose quasi-associative behavior is described in the Equation 4.3 (*vide* Example 4.4).

5. **Symmetry:**

   Symmetry generalizes the associativity and the commutativity for any quantity of evidence to be combined. Observe Equation 4.25, where σ represents the permutation of the arguments.

$$m_1 \oplus m_2 \oplus m_3 \oplus \ldots \oplus m_i = m_{\sigma(1)} \oplus m_{\sigma(2)} \oplus m_{\sigma(3)} \oplus \ldots \oplus m_{\sigma(i)} \tag{4.25}$$

Considering the observations made when we discussed commutativity and of associativity above, symmetry is generically desirable. Our rule models it, with the resource of quasi-associativity (*vide* Equation 4.3 and Example 4.4).

6. **Insensitivity to Vacuousness:**

An evidence aggregation rule must obey the insensitivity to vacuousness property if the combination of an evidence with another that is vacuous results in belief being equal to that of the first evidence, as expressed by the Equation 4.26.

$$m_1 \oplus m_2 = m_1 \text{ , if } m_2(\Theta) = 1 \tag{4.26}$$

In our opinion, this property must not be satisfied, since the vacuous belief mass represents total ignorance or doubt by the source and as the source deserves consideration (because if it did not it should not have being consulted in the first place), its ignorance or doubt must increase the degree of ignorance (our Lateo) in the final result, decreasing concomitantly the belief masses assigned to m_1. This behavior is modeled by our rule. On the other hand, if an aggregation rule has already failed, in the sense of not disposing of an ignorance representation in the result, it is reasonable to expect that the combination with an vacuous evidence vacuous would not change the final result.

4.2.3.3 Conclusions:

The comparison of the proposed rule with the other ones here considered was designed to show not only the imperfections of these rules in modeling some specific situations, but also, the superiority of our rule in the generic modeling of these situations. Respecting the basic premises of the Dempster-Shafer Theory, that is, of the independent sources of evidence as well as the "specialist concept" (that is, the information from all sources consulted deserves to be considered, although it may serve only to aggregate uncertainty to the result), we were unable to model any situation where our rule presented counter-intuitive results, as opposed to what we have shown to happen when other rules are used.

The analysis of the behavior when examined in the light of various mathematical properties confirmed this generic modeling capacity of the proposed rule, when confronted with the argumentations, which constitute the conceptual basis of the extension to the Mathematical Theory of Evidence presented in this work.

5   Case Study

The case study that follows aims to show the applicability of the extension to the Mathematical Theory of Evidence to "real[1]" data from the Brazilian financial market, for the selection of an investment portfolio mix. An interesting thing to notice in this case study is the clarification of the negative aspects that arise from the non-consideration of uncertainty.

In this case study, however, there is no intention to assert either a method for knowledge deduction nor to compare the method used in the portfolio selection with other methodologies, such as Markowitz's Portfolio. We merely intend to show the effects of what happens when two specialists combine their opinions using both Dempster's rule of combination (which does not represents in the result the uncertainty arising from the conflict between the opinions) and the rule proposed in this work (which enforces this representation). As the case is limited to this, the implications of the plausibility and belief functions in the investment strategy were not explored.

The utilization of the data taken from the first half of 2004 for one of the portfolio performance tests had the sole purpose of using data not previously employed by the specialists in their investment mix selections, in order to maintain some coherence with a "real" situation. This study makes no suggestion, implicit or explicit, that the use of the methods described here are able to predict portfolio performance, which assertion would be naive from the point of view of an economist.

5.1   The Problem

A professor, whose income was barely able to cover his monthly expenditures, incurred a large debt due to an unpredictable expense. The only way he saw to pay off the debt was to

---

[1]That is, from the market.

invest his end-of-the-year (2003) bonus plus his holiday pay for a period of six months (the first half of 2004).

At the end of this period, he would pay off the debt from the accumulated interest or pay off what he could and renegotiate the rest.

The indebted professor consulted two specialists regarding the best portfolio mix to choose, being willing to undertake a risk rate proportional to the certainty of the specialists. That is, he would risk his money in a given investment, willing to run the risk of losing his investment if he followed their advice. On the other hand, given the idea to let the indebted professor choose whether or not to accept a greater risk, if the specialists were uncertain about their recommendations, the professor would have the freedom to choose which investment options to select relative to their degrees of uncertainty. Thus the professor could opt to invest in a more conservative investment, where both risk and return potential were low, or in a more aggressive investment, with a greater risk and a larger potential of return.

The investments under consideration (because of features such as the capital available to be invested, the bank where he had an account, and taxes) were:

- DI Fund – a short deadline investment based on post-fixed interests which follows the fluctuations of the official interest rate.

- Savings Account – a classic Brazilian investment that has a return of $TR^2 + 0.5\%$ per month.

- Bonds Fund – a bond based investment fund.

- International Fund – a fund based on the public stock certificates of the Brazilian foreign debt.

From these, the first two are options in which the risk of depreciation of the principal is minimum and the others are options where the chance of the depreciation of the principal is higher.

To facilitate the checking of the results achieved under conditions that would simulate those of the "real" world (that is, for "real" data to be available) the period chosen to test

---

[2]A Brazilian official rate based on the variation of some prices.

the investment was between January and June of 2004. We used data from a "real" Brazilian financial Institution, the Itau Bank, for this test and also for the historic series upon which the specialists based their opinion [39].

As the indebted professor was not a financial expert, both specialists kept in mind that he would not be able to move the investments among the portfolio options nor withdraw any amount during the six-month period.

The first specialist consulted was the manager of the professor's bank account and the second was an economics professor.

The Bank Manager based his opinion on the financial data from the first semesters over the last 36 months, that is, the first semesters of 2001, 2002 and 2003.

The economics professor based his opinion also on the financial data from the first semesters of three years; but, as he did not have up-to-date statistics, he used data from the first semesters of 1998, 1999 and 2000.

Although both specialists based their opinions on the historical data of the performance of the investments, they could have based their opinions on several other factors like intuition, personal background, risk profile of the customer, final amount desired, and economic perspectives.

Obviously, it also must be remembered (as all banks do in their investment contracts) that past performance is not a guarantee of future profits, neither does expert opinion represent a guarantee of profit – both are solely intended to provide evidence upon which the professor could make his final decisions.

The strategy adopted by both specialist for the portfolio mix definition was the following:

- The following factors were kept in mind when deciding on which investments would be in the portfolio:

  1. The accumulated performance of each investment over the period of analysis used by each specialist.

     The assertion that usually the most profitable investment also has more profit variability over time and, consequently, more risk, was considered in the risk degree calculation made to define the percentage of each investment for the portfolio composition.

2. The correlation among the combination of investments.

Regarding this factor, a high[3] positive correlation between a pair of investments would suggest that just one of them, the more profitable one, should be used to compose the portfolio and vice-versa, that is, a high and negative correlation between a pair of investments would suggest an evidence to put both in the port-folio mix and this evidence would be stronger proportionally to the value of this negative correlation.

The purpose of the first factor is to increase the global return of the portfolio, while the second factor has a two-fold function: minimize the risk through composing a portfolio with investments that have a high negative correlation (possibly compensating negative variations in the return of one of them) and increase the return of the portfolio by avoiding making it up of investments that have a high positive correlation, since this kind of correlation suggests a potential advantage in concentrating the resources (that otherwise would be divided among the investments) in the investment option that possesses the greater accumulated return.

- The specialists used the following technique to determine the percentage of capital that would be applied to each investment:

  - They calculated the risk factor of each investment option;

  - They took the inverse of the sum of the risk factors of the investments that com-posed the portfolio;

  - They normalized these results, thus determining the percentage of the total that each investment represented.

The formula used to determine the risk factor, Equation 5.1, keeps in mind the relation of the mean return and the variability of the investment. In this way the risk factor would be higher if the relation between the mean return and the variability were lower. Thus, the investment would have a high risk factor if:

  - It had a high mean return but also a proportionally high variability;

---

[3]The subjective aspect of what is a "high" value is one of the factors that differentiate a finance specialist from a regular person.

   – or it had a low mean return and a proportionally high variability.

Conversely the investment would have a low risk factor if:

   – It had high mean return and a proportionally low variability;

   – or had a low mean return and a proportionally lower variability.

5.2 Mathematical Structure

Putting this problem in the Dempster-Shafer Theory notation, one gets a frame of discernment composed of[4]:

$$\Theta = \{\text{DI Fund, Savings Account, Bonds Fund, Int Fund}\}$$

As it will be shown, the opinions of the 2 specialists created 2 mass functions.

For the risk factor determination the specialists utilized the risk function formula created by Campello de Souza in [13], Equation 5.1.

$$\mathcal{R}(\mathcal{A}) = \left(1 - \frac{\mu(\mathcal{A})}{\mu(\mathcal{A}) + \sigma(\mathcal{A})}\right) \tag{5.1}$$

where:

$\mathcal{R}(\mathcal{A})$ denotes the risk factor of the investment $\mathcal{A}$;

$\mu(\mathcal{A})$ the mean of the return on the investment $\mathcal{A}$;

and $\sigma(\mathcal{A})$ the standard deviation of the return on investment $\mathcal{A}$.

To aid the choice of the portfolio composition even more, the mean, correlation, and accumulated return on the investment were calculated for each analysis period.

5.2.1 Opinion of the first specialist (the Bank Manager)

The Bank Manager analyzed the performance of the investments over the first semesters of the last three years, as can be seen in Table 5.

---

[4]Where "Int Fund" stands for "International Fund".

Based on this data, the Bank Manager calculated the correlation between the investment pairs[5], as it can be seen at Table 6, and the mean return, standard deviation, risk factor and inverse of risk factor, for each of the investments options, these results are shown in Table 7.

Table 5: Historic series of the first semesters from 2001 to 2003 [39]

Period	DI Fund	Accumulated Return	Savings Account	Accumulated Return	Bonds Fund	Accumulated Return	Int Fund	Accumulated Return
Jan/01	0.871	1.009	0.600	1.006	17.047	1.170	4.878	1.049
Feb/01	0.724	1.016	0.638	1.012	-9.086	1.064	2.243	1.072
Mar/01	0.897	1.025	0.537	1.018	-8.476	0.974	3.113	1.106
Apr/01	0.862	1.034	0.673	1.025	5.906	1.031	-0.046	1.105
May/01	0.979	1.044	0.655	1.031	-4.306	0.987	6.301	1.175
Jun/01	0.948	1.054	0.684	1.038	-2.436	0.963	-0.955	1.164
Jan/02	1.187	1.066	0.699	1.046	-3.586	0.928	4.658	1.218
Feb/02	0.963	1.077	0.760	1.054	10.510	1.026	2.374	1.247
Mar/02	1.016	1.088	0.618	1.060	-1.390	1.012	-0.042	1.246
Apr/02	1.076	1.099	0.677	1.067	-0.522	1.006	-1.831	1.223
May/02	-0.103	1.098	0.737	1.075	-1.719	0.989	3.320	1.264
Jun/02	1.044	1.110	0.711	1.083	-8.021	0.910	-2.132	1.237
Jan/03	1.742	1.129	0.863	1.092	-3.158	0.881	4.702	1.295
Feb/03	1.573	1.147	0.990	1.103	-3.227	0.853	6.096	1.374
Mar/03	1.563	1.165	0.914	1.113	7.051	0.913	-1.902	1.348
Apr/03	1.655	1.184	0.880	1.123	6.378	0.971	-7.025	1.253
May/03	1.755	1.205	0.921	1.133	5.358	1.023	2.637	1.286
Jun/03	1.730	1.226	0.967	1.144	-2.108	1.001	-2.997	1.248

Table 6: Correlation between the pairs of investments (1st semesters from 2001 to 2003)

Pair of Investment	Correlation
DI Fund X Savings Account	70.02%
DI Fund X Bonds Fund	12.24%
DI Fund X Int Fund	-23.66%
Bonds Fund X Int Fund	-10.58%

Table 7: Mean Return, Standard Deviation, Risk Factor and Inverse of the Risk Factor (1st semesters from 2001 to 2003)

1^{st}s Semesters 2001 to 2003	DI Fund	Savings Account	Bonds Fund	Int Fund
Mean Return	1.138	0.751	0.234	1.300
Standard Deviation	0.471	0.137	7.030	3.623
Risk	0.293	0.154	0.968	0.736
Inverse of the Risk	3.416	6.498	1.033	1.359

---

[5]The Bank Manager did not need to calculate the correlations among all combinations of pairs of investments because when he calculated the correlation between the DI Fund and the Savings Account he got a high and positive value indicating that, in an economic sense, it would be recommended to choose, between these two investments, selecting the one that exhibited the higher return, not using the other one to compose the portfolio (once a high and positive correlation indicates a lower possibility of one investment minimize the risk of the other). Thus the Savings Account was discarded as an option to be in the portfolio.

The Bank Manager analyzing this information reached the following conclusions:

- As the investments DI Fund and Savings Account had a higher and positive correlation (70,02%) and as the DI Fund presented an accumulated return higher than the Savings Account, the DI Fund was chosen to compose the portfolio.

- A positive correlation of 12,24% between the DI Fund and the Bond Fund suggest that the former had bonds in its own portfolio.

- A negative correlation between the DI Fund and the Int Fund indicated the use of the latter in the composition of the portfolio which was confirmed by its high accumulated return.

- Between the Bonds Fund and the Int Fund there is a limited possibility of mutual variation compensation, evidenced by a low and negative correlation between this pair.

Based on these observations, the DI Fund, the Bonds Fund and the International Fund were chosen to make up the portfolio.

Using the inverse of the risk factor as the sharing criteria to be applied to the investment division, the Bank Manager normalized these values and reached an opinion concerning the share of each investment option in the portfolio. Table 8 shows how the percentage of each investment was raised in proportion to the inverse of its risk factor. These values result in the first mass function:

$$m_1(\{\text{DI Fund}\}) = 0.5881$$
$$m_1(\{\text{Savings Account}\}) = 0$$
$$m_1(\{\text{Bonds Fund}\}) = 0.1779$$
$$m_1(\{\text{Int Fund}\}) = 0.2340$$
$$m_1(\Theta) = 0$$

Table 8: Composition of the portfolio based on the opinion of the Bank Manager

$1^{st} Semesters$ 2001 to 2003	DI Fund	Bonds Fund	Int Fund
Inverse of Risk Factor	3.416	1.033	1.359
Portfolio Share	58.81%	17.79%	23.39%

5.2.2   Opinion of the second specialist (the economics professor)

The economics professor acted in a similar way but analyzed the historic series of the first semesters from 1998 a 2000, shown in Table 9.

Analogous to the Bank Manager, the economics professor based his opinion on the historic series, calculating the correlation between the pairs of investments, in Table 10, and the mean return, standard deviation, risk factor and inverse of risk factor, for each of the investments, in Table 11.

The economics professor observed the following about the correlations between the pairs of investments:

- Similar to how the data had been used by the bank manager, the correlation between the DI Fund and the Savings Account was high and positive (71,51%) justifying the choosing of a higher return option, the DI Fund, instead of the Savings Account.

- The correlation between the DI Fund and the Bonds Fund also was high and positive (47,3%) suggesting the possibility of using the former in the composition of the portfolio instead of the Bonds Fund.

- The correlation between the DI Fund and the Int Fund was positive but too small to eliminate the Int Fund in favor of the DI Fund (which is confirmed by the high accumulated return of the Int Fund).

Considering all this the economics professor constructed a portfolio using only the DI Fund and the Int Fund, dividing the percentage of the capital applied to each fund by the criteria of proportionality to the inverse of the risk factor, Table 12, resulting in the second mass function:

$$m_2(\{\text{DI Fund}\}) = 0.7486$$
$$m_2(\{\text{Savings Account}\}) = 0$$
$$m_2(\{\text{Bonds Fund}\}) = 0$$
$$m_2(\{\text{Int Fund}\}) = 0.2514$$
$$m_2(\Theta) = 0$$

Table 9: Historic series of the first semesters from 1998 to 2000 [39]

Period	DI Fund	Accumulated Return	Savings Account	Accumulated Return	Bonds Fund	Accumulated Return	Int Fund	Accumulated Return
Jan/98	2.307	1.023	1.815	1.018	-9.983	0.900	0.865	1.009
Feb/98	1.809	1.042	1.652	1.035	5.578	0.950	2.645	1.035
Mar/98	1.803	1.060	0.948	1.045	11.560	1.060	3.225	1.069
Apr/98	1.369	1.075	1.404	1.059	3.028	1.092	0.582	1.075
May/98	1.271	1.089	0.974	1.070	-11.710	0.964	-3.294	1.040
Jun/98	1.227	1.102	0.957	1.080	-8.009	0.887	-1.234	1.027
Jan/99	1.819	1.122	1.247	1.093	24.859	1.108	45.033	1.489
Feb/99	2.057	1.145	1.019	1.105	3.749	1.149	9.209	1.626
Mar/99	2.920	1.178	1.334	1.119	29.550	1.489	-10.106	1.462
Apr/99	2.006	1.202	1.667	1.138	3.572	1.542	-0.302	1.457
May/99	1.641	1.222	1.112	1.151	-0.939	1.528	2.521	1.494
Jun/99	1.298	1.238	1.079	1.163	7.973	1.649	5.491	1.576
Jan/00	1.085	1.251	0.801	1.172	-1.824	1.619	-0.240	1.572
Feb/00	1.084	1.265	0.716	1.181	6.724	1.728	1.722	1.599
Mar/00	1.087	1.278	0.734	1.189	-1.253	1.707	1.377	1.621
Apr/00	0.966	1.291	0.725	1.198	-7.812	1.573	-0.110	1.620
May/00	1.113	1.305	0.631	1.206	1.317	1.594	-0.261	1.615
Jun/00	1.027	1.319	0.750	1.215	11.307	1.774	4.751	1.692

Table 10: Correlation of the investment pairs (1st semesters from 1998 to 2000)

Pair of Investment	Correlation
DI Fund X Saving Account	71.51%
DI Fund X Bonds Fund	47.31%
DI Fund X Int Fund	1.41%
Bonds Fund X Int Fund	39.67%

Table 11: Mean Return, Standard Deviation, Risk Factor and Inverse of the Risk Factor (1st semesters from 1998 to 2000)

1st Semesters 2001 to 2003	DI Fund	Savings Account	Bonds Fund	Int Fund
Mean Return	1.138	0.751	0.234	1.300
Standard Deviation	0.471	0.137	7.030	3.623
Risk Factor	0.293	0.154	0.968	0.736
Inverse of Risk Factor	3.416	6.498	1.033	1.359

Table 12: Participation of each investment in the portfolio based on the opinion of the economics professor

1st Semesters 1998 to 2000	DI Fund	Int Fund
Inverse of the Risk Factor	3.898	1.309
Percentage of the Portfolio	74.86%	25.14%

Now we will set up other portfolios based on the combination of the specialists' opinions within the frameworks of Dempster's Rule and the new rule proposed in the present work.

5.2.3   Combining using Dempster's Rule

Combining both specialists' opinions (m_1 e m_2) using Dempster's Rule, Table 13, one gets, after the normalization, the following values of combined belief masses:

$$m_3(\{\text{DI Fund}\}) = 0.8821$$

$$m_3(\{\text{Savings Account}\}) = 0$$

$$m_3(\{\text{Bonds Fund}\}) = 0$$

$$m_3(\{\text{Int Fund}\}) = 0.1179$$

$$m_3(\Theta) = 0$$

Table 13: Combination of the opinions using Dempster's Rule

	$m_1(\{\text{DI Fund}\})$ $= 0.5881$	$m_1(\{\text{Savings Account}\})$ $= 0$	$m_1(\{\text{Bonds Fund}\})$ $= 0.1779$	$m_1(\{\text{Int Fund}\})$ $= 0.2340$	$m_1(\Theta)$ $= 0$
$m_2(\{\text{DI Fund}\}) = 0.7486$	0.4403	0	0	0	0
$m_2(\{\text{Savings Account}\}) = 0$	0	0	0	0	0
$m_2(\{\text{Bonds Fund}\}) = 0$	0	0	0	0	0
$m_2(\{\text{Int Fund}\}) = 0.2514$	0	0	0	0.0588	0
$m_2(\Theta) = 0$	0	0	0	0	0
	$\sum = 0.4403$	$\sum = 0$	$\sum = 0$	$\sum = 0.0588$	$\sum = 0$

5.2.4   Combining by the new rule proposed in this work

Combining using the new rule one gets the following values for the combined masses of belief:

$$m_3(\{\text{DI Fund}\}) = 0.6776$$

$$m_3(\{\text{Savings Account}\}) = 0$$

$$m_3(\{\text{Bonds Fund}\}) = 0$$

$$m_3(\{\text{Int Fund}\}) = 0.0905$$

$$\Lambda = 0.2319$$

In opposition to the combination by Dempster's Rule, with the combination using the new rule, Lateo makes it explicit in the result that in the combination of the opinions there was uncertainty (in this particular case the uncertainty came from the conflict between the opinions of specialists). Thus the indebted professor is given the option to assign this mass of uncertainty to the investment, or investments, that best match the risk and return potentials that he has looked at. Therefore, the professor can opt for any distribution combination of the Lateo among the options from the frame of discernment (even the Savings Account or the Bonds Fund which were not included in the portfolio assembled by the specialists).

To simplify the illustration of the behavior of Lateo's assignment, just two of the possible attribution options will be explored: the assignment of all value of Lateo to the DI Fund (which represents the adoption of a conservative profile) and the attribution of all of it to the Int Fund (meaning a more aggressive investment profile).

The beliefs masses for the first combination alternative, the conservative profile, would suggest a portfolio with the composition:

$$m_3(\{\text{DI Fund}\}) + \Lambda = 0.9095$$
$$m_3(\{\text{Savings Account}\}) = 0$$
$$m_3(\{\text{Bonds Fund}\}) = 0$$
$$m_3(\{\text{Int Fund}\}) = 0.0905$$

The aggressive profile, represented by the assignment of all the value of the Lateo to the Int Fund would result in the following beliefs masses and portfolio composition:

$$m_3(\{\text{DI Fund}\}) = 0.6776$$
$$m_3(\{\text{Savings Account}\}) = 0$$
$$m_3(\{\text{Bonds Fund}\}) = 0$$
$$m_3(\{\text{Int Fund}\}) + \Lambda = 0.3224$$

5.3   Performance of the Portfolios

Based on the values previously determined by the composition of the portfolios, performance will be calculated using historic data from the first semester of 2004, Table 14, date determined by the indebted professor as his investment period.

Table 14: Historic series of the first semester of 2004 [39]

Period	DI Fund	Accumulated Return	Savings Account	Accumulated Return	Bonds Fund	Accumulated Return	Int Fund	Accumulated Return
Jan/04	0.998	1.010	0.691	1.007	-0.524	0.995	0.435	1.004
Feb/04	0.786	1.018	0.629	1.013	1.266	1.007	-3.106	0.973
Mar/04	0.993	1.028	0.546	1.019	3.256	1.040	1.372	0.987
Apr/04	0.767	1.036	0.679	1.026	-10.023	0.936	-6.227	0.925
May/04	0.890	1.045	0.588	1.032	-0.448	0.932	3.103	0.954
Jun/04	0.923	1.055	0.660	1.039	6.489	0.992	3.974	0.992

5.3.1   Performance of the portfolio designed by the Bank Manager

The Table 15 shows the performance of the portfolio designed using only the opinion of the Bank Manager.

Table 15: Performance of the portfolio designed by the Bank Manager

	DI Fund	Savings Account	Bonds Fund	Int Fund	
Portfolio Composition	0.588	0	0.178	0.234	
Return Jan/04 to Jun/04	1.055	1.039	0.992	0.992	
Investment Return	0.620	0	0.177	0.232	
Portfolio Return					1.029

5.3.2   Performance of the portfolio designed by the economics professor

Table 16 shows the performance of the portfolio designed by the economics professor.

Table 16: Performance of the portfolio designed by the economics professor

	DI Fund	Savings Account	Bonds Fund	Int Fund	
Portfolio Composition	0.7486	0	0	0.2514	
Return Jan/04 to Jun/04	1.055	1.039	0.992	0.992	
Investment Return	0.790	0	0	0.249	
Portfolio Result					1.039

5.3.3 Performance of the portfolio assembled using Dempster's Rule

Utilizing the portfolio suggested by the masses of belief reached by the combination using Dempster's Rule, one gets the results shown in Table 17.

Table 17: Performance of the portfolio suggested by the Dempster's Rule

	DI Fund	Savings Account	Bonds Fund	Int Fund	
Portfolio Composition	0.882	0	0	0.118	
Return Jan/04 to Jun/04	1.055	1.039	0.992	0.992	
Investment Return	0.930	0	0	0.117	
Portfolio Return					1.047

5.3.4 Performance of the portfolio when the new rule is applied

Utilizing the new rule proposed in this work, considering the conservative profile, the beliefs masses would result in a portfolio with the performance shown in Table 18.

When the new rule is used with the aggressive profile, the performance of the resulted portfolio would be that shown in 19.

Table 18: Performance of the portfolio using the new rule of combination (conservative profile)

	DI Fund	Savings Account	Bonds Fund	Int Fund	
Portfolio Composition	0.9095	0	0	0.0905	
Return Jan/04 to Jun/04	1.055	1.039	0.992	0.992	
Investment Return	0.959	0	0	0.090	
Portfolio Result					1.049

Table 19: Performance of the portfolio using the new rule with the aggressive profile

	DI Fund	Savings Account	Bonds Fund	Int Fund	
Portfolio Composition	0.678	0	0	0.322	
Return Jan/04 to Jun/04	1.055	1.039	0.992	0.992	
Investment Return	0.715	0	0	0.320	
Portfolio Return					1.034

5.3.5   Others performances of the portfolios

Projecting from the situations we have set up, we can calculate the performances of the portfolios dealt with by combining the opinions of the specialists using Dempster's Rule and the performance of both portfolio profiles employing the new rule, using the data of the seven first semesters chosen for this case study, that is, those between 1998 and 2004. Tables 20 and 21 show these results.

Table 20: Comparison year by year of the portfolios' performance

Period	New Rule (conservative)	Better performance than Dempster's?	New Rule (Aggressive)	Better performance than Dempster's?	Dempster's Rule
1st semester 1998	1.095	Yes	1.078	Yes	1.093
1st semester 1999	1.161	No	1.256	Yes	1.172
1st semester 2000	1.066	Equal	1.068	Yes	1.066
1st semester 2001	1.064	No	1.089	Yes	1.067
1st semester 2002	1.054	Equal	1.056	Yes	1.054
1st semester 2003	1.096	Yes	1.073	No	1.093
1st semester 2004	1.049	Yes	1.034	No	1.047

Table 21: Comparison of the accumulated return over 7 years

Rule	Accumulated Result
New Rule (conservative profile)	7.46%
New Rule (aggressive profile)	8.46%
Dempster's Rule	7.59%

5.3.6   Some points about the performances of the portfolios

With respect to the portfolio performance during the first semester of 2004:

- The portfolio suggested by the combination of the specialists' opinions using Dempster's Rule resulted in a return better than the best of the specialists' portfolio.

- The portfolio suggested by the new rule, considering a conservative profile, resulted in a return higher than the portfolios suggested by the specialists' opinions and by the combination of the opinions using the Dempster's Rule.

- The portfolio assembled on the basis on the new rule with an aggressive profile resulted in a return lower than the one suggested by Dempster's Rule; and using the new rule with a conservative profile, which could be expected, since a well constructed conservative portfolio tends to give a better mean return than one from an aggressive portfolio in the short run.

Regarding the performance over the period of seven semesters:

- Both the portfolio arising out of the combination using Dempster's Rule and the ones assembled using the new rule were successful in preserving the principal, even when applied over a period of seven years.

- The two portfolios (equivalent to the conservative and aggressive profiles) arising from the new rule showed a superior performance over more semesters than the portfolio suggested by Dempster's Rule. In the case of the portfolio associated with the conservative profile, the performance was better than or equal to the one derived from Dempster's Rule in 5 of the 7 semesters. The portfolio with the aggressive profile, using the new rule, was better than when using Dempster's Rule in 5 out of 7 semesters.

- As it would be expected (in an economic sense) the portfolio based on the aggressive profile tends to result in better performance over the long term, which was confirmed by the accumulated return of the portfolios over the 7 semester period, when the performance of the portfolio constituted through the use of the new rule, with an aggressive profile, was superior to the other portfolios.

5.4   Conclusions from this case study

Secondary results:

- Better performance of the portfolio using the proposed rule in the conservative profile in the period of the first semester of 2004;

- Better performance of both portfolios using the proposed rule in the number of semesters that they achieved higher returns than the investment portfolios using Dempster's Rule.

We consider these results secondary ones because it is not our intention , in this work, to find a way to predict investment performance.

Primarily this case study shows some relevant aspects that confirm the importance of the extension made to the Mathematical Theory of Evidence:

- The possibility of its practical application in situations where the information is scarce and conflicting.

- The use of Lateo to clarify for the user that discord exists among the specialists, a conflict that is hidden in Dempster's Rule.

- The representation of uncertainty provided by the Lateo, makes the use of uncertainty possible as a strategic asset, the manipulation of which can result in significant practical differentials. Therefore, it is not adequate to simply redistribute it among bodies of evidence that contain some belief, as does Dempster's Rule.

6 Conclusions

The Dempster-Shafer's Theory, or Mathematical Theory of Evidence, has been being broadly used in the more diverse branches of human knowledge, as a formalism for the representation and combination of the knowledge subject to subjective uncertainty. With the evidence combination rules used up to the present, this theory has two main deficiencies:

- A counter-intuitive behavior when the bodies of evidence to be combined contain belief concentration in disjoint elements and has a common element with low values of belief attributed to it;

- the lack, in the result, of an intrinsic and generic mechanism of representation of uncertainty (arising from ignorance or conflict between the evidence).

Through the adoption of the new combination rule proposed in this work, it is possible to solve these two deficiencies, extending the expressive power of this theory and, consequently, its range of applications. This rule corrects counter-intuitive behavior and incorporates, in the result, uncertainty arising from ignorance or conflict in the evidence. That happens by diminishing the beliefs proportionally to the weight of conflict among the evidence and by assigning the remaining belief to the environment, rather than to the common element. With the proposed rule, it becomes possible to combine evidence that owe most of their attributed beliefs to disjoint hypotheses, without the side effect of counter-intuitive behavior. It is also possible the utilization of bodies of evidence that possesses high values of conflict, making useful evidence that otherwise would be discarded, useless.

With the solution of these problems and the implementation of a measure of the ignorance and conflict in the evidence (consequence of our rule), the Lateo, we arrive not just at the aforesaid extension of the Dempster-Shafer's Theory, but also at a formalism able to eliminate the dichotomy of treatment, both formally and conceptually, between traditional probability and the numbers arising from the evidence. In our formalism, numeric and conceptual relations between the two kinds of beliefs become clear in the numeric results obtained, as well as in the connections of the results to "rational bets" and "rational decisions".

6.1   Practical and Conceptual Implications

6.1.1   An Extension to the Mathematical Theory of Evidence

The rules of the combination of evidence used up to the present time have merit when employed in the modeling of specific situations, but fail when we try to utilize them in the generic modeling of situations[1], lacking the support of a conceptual basis that justifies their behavior in these situations.

The adoption of our rule of evidence combination enables the solution of two structural problems contained in the Dempster-Shafer Theory: the absence of an intrinsic way of representing uncertainty and conflict between evidence in the result, and the counter-intuitive behavior of the combination rules when subject to certain conditions in the combination of evidence. The resolution of these problems allows for the construction of an entire conceptual basis that extends the application range of the theory to such a point as to allow the generic treatment of both the subjective and objective uncertainties, resulting in the "Extension to the Mathematical Theory of Evidence".

With our rule we managed to show the viability of the application of normalization without getting counter-intuitive behavior. This fact, more than just being a mathematical detail, results in several important conceptual implications. Among them, it endows the operation of the combination of evidence with the quality of an additive measure, just as always has been thought, in an epistemic sense, about the probabilities (that is, always resulting in the division of 100% of it among the sample space), as well as bringing down several ideas conceived along these decades of development on Dempster-Shafer's Theory, that attributed the cause of the counter-intuitive behavior to the normalization operation (without however, solving the conceptual problem of the non-additivity of the results in the absence

---

[1]That is, the rules are useful in practical applications as long as they are not be subjected to the conditions that lead to the counter-intuitive behaviors, incurring in this kind of behavior if the precautions to avoid the submission of these critical conditions are not taken. Once taken, these precautions limit the range of applications and modeling power of the theory.

of the normalization operation). Several authors, among them Smets, in [50], and Zadeh[2], in [70], criticized the normalization and attributed the occurrence of the counter-intuitive behavior to it. Others, like Clarke in his critical reply to a Smets' article in [50], recognized the absence of the representation in the result of the ignorance and conflict between evidence and suggested the adoption of non-justifiable (in an epistemic sense) principles as the solution for the counter-intuitive behavior (such as the averaging of the more credited beliefs of the conflicting hypotheses). On the other hand, authors like Gerhard Paass (in [50]) and Klir (in [25]), have moved towards a correct solution, although not reaching it, when they stated the implausibility of every combination of bodies of evidence always indicate no measure of contradiction in the results. Klir, for example, reports in [25] that an important aspect of all mathematical theory that conceptualizes uncertainty of any kind, it is its capacity to quantify the uncertainty involved.

6.1.2 The interpretation of uncertainties provided by the Extension to the Mathematical Theory of Evidence

Even though the question of counter-intuitive behavior has been solved, two of the afore-mentioned inquiries still remain unanswered:

- What do the probabilities expressed by the results of the combination of evidence mean?

- How do these "probabilities" relate to the "classic probabilities"?

6.1.2.1 Origin of the dichotomy between "classic probability" and "rational belief"

Before we explain how the "Extension to the Mathematical Theory of Evidence" relates to classic probability and to probability arising from the combination of evidence, it is important to consider the historical context in which these two kinds of beliefs[3] became separated:

According to Shafer, classic probability describes, in a first instance, a special and unusual situation where frequency and the belief are unified. This special situation is the chance in classical game theory involving a sequence of experiments (like successive throws of dice, the

---

[2]Cited in [51] as the first author to exemplify the counter-intuitive behavior of the Dempster-Shafer's Rule, in his work [70].

[3]Historic material extracted from the work of Shafer, [43].

shuffling of a deck of cards, and flips of a coin) that result in well-known frequencies after a large number of repetitions.

Probabilists have been debating frequency and "rational belief" approaches for more than a 100 years. Frequencies in multiple repetitions and rational belief were regarded as complementary aspects of probability during the 18th century and middle of the 19th, but by the middle to the end of the 19th century, rational belief was challenged by the empiricists who considered it too metaphysical. Frequency in large number repetitions was the only foundation they admitted for Probability Theory.

The frequentism of the 19th century still dominates thinking about probability, but this thought was challenged in the 20th century by a resurgent subjectivism led by Bruno de Finetti and L. J. Savage [37]. However, modern subjectivists, or Bayesians, are as empirical as the frequentists of the 18th century, since they share the same frequentist disdain for degrees of belief.

6.1.2.2 Relations until now accepted between the belief arising from evidence and classic probability

Still paraphrasing Shafer in [43], we have his answer to some of the questionings that had been raised regarding the significance and utility of his theory, as well as between the relation of its results to classic probability:

- Regarding the question of why to use the formalisms based on belief functions in detriment to traditional probability, Shafer answers that the main reason is not due to a preference for belief functions over traditional probability, but its usefulness in cases where we cannot manage to build traditional probabilities convincingly.

- In what concerns the association between the Belief and Plausibility Functions with lower and upper probability borders, respectively, Shafer considers this matter disappointing, since he had assigned these names to the functions in order to distinguish them from the probabilities borders.

Formally, any belief function, $\mathcal{B}el$, in a frame of discernment, Θ, is a lower probability function. If we consider p as the family of all the probability distributions P in Θ, we would have for each subset $\mathcal{A}$ of Θ:

$$\mathcal{B}el(\mathcal{A}) = inf\{P(\mathcal{A})|P \in p\} \tag{6.1}$$

But, what interpretation should be given to the probabilities expressed by the probabilities distributions of p?

For Shafer, in the absence of a frequentist interpretation, we must follow what Isaac Levi wrote and say that p is a class of belief distribution among which we cannot decide [43]. His conclusion is that in this case the family p is purely a mathematical construct, without any conceptual significance.

6.1.2.3   Attributes of the Lateo

The implementation of the Lateo in the result, that is, a way to represent the ignorance and the conflict between evidence, leads to several considerations, since its value indicates a series of attributes:

1. how far we are from the classic probability numeric values;

2. how much we can trust in the obtained results, in order to make the decision;

3. or as an indication of the degree of uncertainty of the sources; or a indication of the degree of knowledge of the sources with respect to solving the matter at hand.

We are going to explain each one of these attributes:

Attribute 1: in the case of a non-null Lateo, its value indicates a quantity of conflict between the evidence, or a quantity of ignorance regarding the beliefs attributed to the hypotheses. This value could be associated to any element of the frame of discernment, in case the evidence considered did not introduce any conflict or ignorance *a priori*. Thus, numerically, the Lateo corresponds to a belief value that could not be attributed to any element of the frame of discernment, moving us away from classic probability in the sense that it does not admit belief masses that cannot be attributed to singletons. Numerically, in the form of classic probability, the Lateo represents probability values that could be, in fact (that is, if we had obtained information without conflict or ignorance), divided among the elements of the frame, even among the elements that did not previously receive belief.

With that, from a formal point of view, we become able to consider the belief and plausibility functions as lower and upper limits, respectively, of classic probability bounds; however, the larger the Lateo value, the broader and consequently more imprecise will become this interval, losing its practical utility for high values.

Attribute 2: as explained above, the Lateo represents a belief value that can be associated to any element of the frame of discernment, thus, its value indicates if and how much we can trust in the beliefs attributed to the focal elements to the decision-making. Vide Example 4.6. Note that the magnitude of the Lateo indicates the degree of confidence that we can have in the results; a null Lateo indicates that there was no conflict or ignorance in the sources consulted, that is, it indicates a high reliability in the results obtained; a small value of Lateo indicates a small degree of conflict or ignorance, resulting in a smaller level of reliability; the larger the value of Lateo, the less confidence we can attribute to the belief values assigned to the elements.

Attribute 3: the Lateo is directly associated to the degree of uncertainty derived from the amount of ignorance or conflict between the evidence. A little subtler, is the association that can be made between the Lateo and the capacity of the sources consulted to solve the matter at hand. This latter association allows the Lateo to act like an indicator to determine if the sources contain enough information to solve the matter under study, and indicating the need to consult additional or alternative sources.

6.1.3   Evaluation of the proposed new rule of evidence combination when compared with other combination rules and submitted to some mathematical properties

As it can be seen in 4.2.3, the comparison of our rule with other rules considered here shows not only the imperfections in the other rules in modeling some specific situations, but also, the superiority of our rule in the generic modeling of these situations[4]. No situation was found where our rule exhibited a counter-intuitive result, in opposed to what was found in the other rules[5].

---

[4]Respected Dempster-Shafer's Theory basic premises of independence of the evidences sources and the specialist's concept (that is: the information of all consulted sources deserves to be considered, even though it just aggregates uncertainty to the result).

[5]In particular, can be shown that our rule of combination is able to solve the situations of epistemic incoherence exemplified by Walley in [60], which constitute his main objection to the Mathematical Theory

There are some secondary results worthy of being cited:

- Clarification of the fact that the normalization operation itself is not responsible for the counter-intuitive behavior of the combination rules (the opposite is sustained by Yager and accepted in a generic way by the community).

- The affirmation that contradicts the one sustained by Inagaki that any combination of rules of evidence can be expressed by his equations.

- The superior performance of our rule even in the case of hypotheses with non-disjoint elements, since, whatever the case, it will devalue the beliefs attributed to the hypotheses proportionally to the weight of conflict among them, allowing the combination even when there is a high weight of conflict.

- Our rule is the only one that allows for the modeling of a paradigm change caused by the accumulation of disruptive evidences (vide Section 4.2.2.1), because it accepts this kind of evidence and adds to the result, through the Lateo, the uncertainty representation caused by its accumulation, therefore characterizing the paradigm change phase. And as the inputs that agree with the new paradigm increase, the Lateo, representing the uncertainty attributed to the environment, will decrease, characterizing the new paradigm sedimentation as the most acceptable solution, reproducing epistemic reasoning.

Behavior analysis, when submitted to the mathematical properties of idempotence, commutativity, continuity, associativity, symmetry, and insensitivity to vacuousness, has the purpose of confirming the generic modeling capability of our rule along with the several conceptual argumentations we have defended. Such argumentations constitute the conceptual basis on which our "Extension to the Mathematical Theory of Evidence" is supported. It must be stressed that none of the other rules cover in generic form (which ours does) assistance to all of the conceptual questionings exhibited.

---

of Evidence.

6.2   Final Comments

We have attempted to show that the adoption of our rule of evidence combination, jointly with an adequate conceptual framework, extends Dempster-Shafer's Theory, resulting in what we named an "Extension to the Mathematical Theory of Evidence", which enables:

- An increase in the original applications range of the Dempster-Shafer Theory, increasing substantially its power of expression;

- a solution for the structural problems of this theory, namely, counter-intuitive behavior and the lack of an intrinsic representation for conflict or ignorance of the sources in the result;

- the elimination of ideas, conceived over decades, that the normalization operation or the additivity of the combined beliefs is the cause of the counter-intuitive behavior of Dempster's Rule;

- the capability for uniform treatment, both formally and conceptually, of objective and subjective uncertainties, eliminating a dichotomy existing up to the present;

- the automatic modeling, through the Lateo, of the several attributes described in Section 6.1.2.3, which were total or partially masked by the combination rules used up to the present, even though they have been extremely useful for the pragmatic application of a theory of this nature.

References

[1] Thomas Bayes. *An Essay Toward Solving a Problem in the Doctrine of Chances*, volume 53. 1763. Reprinted in *Facsimiles of two papers by Bayes*, Hafner Publishing Company, New York, 1963.

[2] N.D. Belnap. A useful four-valued logic. In J.M. Dunn and G. Epstein, editors, *Modern Uses of Multiple-Valued Logics*. D.Reidel Pub. Co., 1977.

[3] Michael Berg. *Becoming Like God*. Kabbalah Publishing, Los Angeles, 2004. ISBN 1-57189-242-7.

[4] Guilherme Bittencourt. Tutorial em inteligência artificial. http://www.das.ufsc.br/, accessed in 01/05/2003.

[5] P. Bonissone. Plausible reasoning. In S. C. Shapiro, D. Eckroth, and G.A. Valassi, editors, *Encyclopedia of Artificial Intelligence*, pages 854–863. John Wiley & Sons Inc, New York, USA, 1991.

[6] George E. P. Box. Science and statistics. *Journal of the American Statistical Association*, (71):791–799, 1976.

[7] Fabio Campos. *Uma Extensão à Teoria Matemática da Evidência*. PhD thesis, Centro de Informática - Universidade Federal de Pernambuco, 2004.

[8] Fabio Campos and Fernando Campello de Souza. Extending dempster-shafer theory to overcome counter intuitive results. In *IEEE-NLP-KE-05*, Wuhan, China, 2005.

[9] Fabio Campos and Sergio Cavalcante. An extended approach for Dempster-Shafer theory. In *IEEE-IRI-03*, Las Vegas, USA, 2003.

[10] Fabio Campos and Sergio Cavalcante. A method for knowledge representation with automatic uncertainty embodiment. In *IEEE-NLP-KE-03*, Beijing, China, 2003.

[11] Vijay Chandru and John N. Hooker. *Optimization Methods for Logical Inference*. Series in Discrete Mathematics and Optimization. John Wiley & Sons, 1999. ISBN 0-471-57035-4.

[12] R. Courant. *Diferential and Integral Calculus*, volume I. Blackie & Son, London, 26 edition, 1959.

[13] Fernando Menezes Campello de Souza. *Decisões Racionais em Situações de Incerteza*. Editora Universitária, Recife, PE, Brazil, 2002.

[14] Fernando Menezes Campello de Souza. Sistemas probabilísticos. Unpublished book, UFPE, Brazil, 2004.

[15] Fernando Menezes Campello de Souza, Bruno Campello de Souza, and Alexandre Stamford da Silva. *Elementos da Pesquisa Científica em Medicina*. Editora Universitária, Recife, PE, Brazil, first edition edition, July 2002.

[16] A. P. Dempster. Upper and lower probabilities induced by a multivalued mapping. *The Annals of Mathematical Statistics*, 38(2):325–339, April 1967.

[17] Gilene do Espírito Santo Borges. Sgmoo: Sistema gestor de métodos orientados a objetos baseado em conhecimento. Master's thesis, Departamento de Ciência da Computação - Universidade de Brasília, 1998.

[18] D. Dubois and H. Prade. *Fuzzy Sets and Systems: Theory and Applications*. Academic Press, 1980.

[19] D. Dubois and H. Prade. *Possibility Theory - An Approach to the Computerized Processing of Uncertainty*. Academic Press, 1988.

[20] D. Dubois and H. Prade. On the combination of evidence in various mathematical frameworks. In J. Flamm and T. Luisi, editors, *Reliability Data Collection and Analysis*, pages 213–241, Brussels, 1992. ECSC, EEC, EAFC.

[21] Scott Ferson, Vladik Kreinovich, Lev Ginzburg, Davis S. Myers, and Kari Sentz. Constructing probability boxes and Dempster-Shafer structures. *Sand Report*, January 2003. Unlimited Release.

[22] Wilhelmiina Hämäläinen. Search for an ideal reasoning system: specification of the minimal requirements. From the website of the Department of Computer Science, University of Joensuu, Finland, 2004.

[23] J. C. Helton. Uncertainty and sensitivity analysis in the presence of stochastic and subjective uncertainty. *Journal of Statistical Computation and Simulation*, 57:3–76, 1997.

[24] A.V. Joshi, S.C. Shasrabudhe, and K. Shankar. Sensivity of combination schemes under conflicting conditions and a new method. In J. Wainer, A. Carvalho, et al., editors, *Advances in Artificial Inteligence*, Lectures Notes in Computer Science, pages 39–47. Springer-Verlag, 1995.

[25] George J. Klir. Measures of uncertainty in the Dempster-Shafer theory of evidence. In Ronald R. Yager, Mario Fedrizzi, and Janusz Kacprzyk, editors, *Advances in the Dempster-Shafer Theory of Evidence*. John Wiley & Sons, Inc, Nova York, USA, 1994. ISBN 0-471-55248-8.

[26] George J. Klir and T.A. Folger. *Fuzzy sets, uncertainty, and information*. Prentice-Hall, Englewood Cliffs, NJ, USA, 1988.

[27] J. Kohlas and P.-A. Monney. Representation of evidence by hints. In Ronald R. Yager, Mario Fedrizzi, and Janusz Kacprzyk, editors, *Advances in the Dempster-Shafer Theory of Evidence*. John Wiley & Sons, Inc, Nova York, USA, 1994. ISBN 0-471-55248-8.

[28] H.E. Kyburg, Jr. The reference class. *Philosophy of Science*, 50:374–397, 1983.

[29] Victor Lesser. Lecture on theory of evidence – slides of the lecture at the cmpsci cached by google. http://www.google.com, 11/12/02.

[30] Gertudres Coelho Nadler Lins and Fernando Menezes Campello de Souza. A protocol for the elicitation of prior distribution. *2nd International Symposium on Imprecise Probabilities and Their Applications*, 2001.

[31] E.H. Mamdani, John Bigham, and Flash Sheridan. Introduction. In Philippe Smets, E.H. Mamdani, Didier Dubois, and Henri Prade, editors, *Non-Standard Logics for Automated Reasoning*. Academic Press, San Diego CA, USA, 1988. ISBN: 0-12-649520-3.

[32] J. McCarthy. Circumscription - a form of non-monotonic reasoning. *Artificial Intelligence*, 13(1-2):27–39, April 1980.

[33] Hung T. Nguyen and Elbert A. Walker. On decision making using belief functions. In Ronald R. Yager, Mario Fedrizzi, and Janusz Kacprzyk, editors, *Advances in the Dempster-Shafer Theory of Evidence*. John Wiley & Sons, Inc, Nova York, USA, 1994. ISBN 0-471-55248-8.

[34] Gerhard Paass. Probabilistic logic. In Philippe Smets, E.H. Mamdani, Didier Dubois, and Henri Prade, editors, *Non-Standard Logics for Automated Reasoning*. Academic Press, San Diego CA, USA, 1988. ISBN: 0-12-649520-3.

[35] Z. Pawlak. Rough sets: a new approach to vagueness. In Lotfi A. Zadeh and J. Kacprzyk, editors, *Fuzzy Logic for the Management of Uncertainty*. John Wiley & Sons Inc, New York, USA, 1992.

[36] J. Pearl. *Probabilistic Reasoning in Intelligent Systems: Networks of Plausible Inference*. Morgan Kaufmann Publishers Inc, San Mateo, CA, USA, 1987.

[37] Jan Von Plato. *Creating Modern Probability*. Cambridge University Press, UK, 1998. ISBN: 0-521-59735-8.

[38] R. Reiter. A logic for default reasoning. *Artificial Intelligence*, 13(1-2):81–132, April 1980.

[39] Banco Itaú S.A. *Séries históricas de investimentos fornecidas pelo Banco Itaú*. Banco Itaú Website. http://www.itau.com.br, accessed in 08/20/2004.

[40] S. Sandri. Local propagation of information on directed markov trees. In B. Bouchon-Meunier et al., editors, *Uncertainty in Intelligent Systems*. Elsevier Science Publishers, 1993.

[41] S. Sandri, D. Dubois, and H. Prade. Elicitation, pooling and assessement of expert judgments using possibility theory. *IEEE Transactions on Fuzzy Systems*, 1995.

[42] Kari Sentz and Scott Ferson. Combination of evidence in Dempster-Shafer theory. *Sand Report*, April 2002. Unlimited Release.

[43] Glenn Shafer. Response to the discussion of belief function. http://www.glennshafer.com/index.html, accessed in 06/21/2002.

[44] Glenn Shafer. What is probability. http://www.glennshafer.com/index.html accessed in 10/19/2002.

[45] Glenn Shafer. *A Mathematical Theory of Evidence*. Princeton University Press, 1976. ISBN: 0-691-08175-1.

[46] Glenn Shafer. Perspectives on the theory and practice of belief functions. *International Journal of Approximate Reasoning*, 4(5-6):323–362, 1990.

[47] Glenn Shafer. *The Art of Causal Conjecture*. The MIT Press, London, UK, 1996. ISBN: 0-262-19368-X.

[48] Glenn Shafer. *Probabilistic Expert Systems*. CBMS-NSF Regional Conference Series in Applied Mathematics. SIAM - Society for Industrial and Applied Mathematics, Philadelphia, USA, 1996. ISBN: 0-89871-373-0.

[49] P. P. Shenoy. Using Dempster-Shafer's belief-function theory in expert systems. In Ronald R. Yager, Mario Fedrizzi, and Janusz Kacprzyk, editors, *Advances in the Dempster-Shafer Theory of Evidence*. John Wiley & Sons, Inc, Nova York, USA, 1994. ISBN 0-471-55248-8.

[50] Philippe Smets. Belief functions. In Philippe Smets, E.H. Mamdani, Didier Dubois, and Henri Prade, editors, *Non-Standard Logics for Automated Reasoning*. Academic Press, San Diego CA, USA, 1988. ISBN: 0-12-649520-3.

[51] Philippe Smets, E.H. Mamdani, Didier Dubois, and Henri Prade, editors. *Non-Standard Logics for Automated Reasoning*. Academic Press, San Diego CA, USA, 1988. ISBN: 0-12-649520-3.

[52] R. Stein. The Dempster-Shafer theory of evidential reasoning. *Artificial Intelligence Expert*, August 1993.

[53] William N. Stephens. *Hypotheses and Evidence*. Thomas Y. Crowell Company, 1968. Florida Atlantic University, Library of Congress Catalog Card Number: 68-13387.

[54] Thomas M. Strat. Decision analysis using belief functions. In Ronald R. Yager, Mario Fedrizzi, and Janusz Kacprzyk, editors, *Advances in the Dempster-Shafer Theory of Evidence*. John Wiley & Sons, Inc, Nova York, USA, 1994. ISBN 0-471-55248-8.

[55] K. Tanaka and George J. Klir. A design condition for incorporating human judgement into monitoring systems. *Reliability Engineering and System Safety*, 65(3):251–258, 1999.

[56] A. Trusov. *An Introduction to the Theory of Evidence*. Foreign Language Publishing House Moscow, Moscow, 1962.

[57] Joaquim Quinteiro Uchoa, Sônia Maria Panotim, and Maria do Carmo Nicoletti. Elementos da teoria da evidência de Dempster-Shafer – tutorial do departamento de computação da universidade federal de são carlos. http://www.dc.ufscar.br, accessed in 11/18/02.

[58] Peter Walley. *Statistical Reasoning with Imprecise Probabilities*. Chapman and Hall, London, 1991.

[59] Peter Walley. Inferences from multinomial data: Learning about a bag of marbles. *Royal Statistical Society*, B(1):3–57, 1996.

[60] Peter Walley. Measures of uncertainty in expert systems. *Artificial Intelligence*, 83(1):1–58, 1996.

[61] Peter Walley. Coherent upper and lower previsions. *Imprecise Probabilities Project Website*, 1997. http://ensmain.rug.ac.be/ ipp, accessed in 10/20/2002.

[62] Peter Walley. Towards a unified theory of imprecise probability. *Int. J. Approx. Reasoning*, 24(2-3):125–148, 2000.

[63] Peter Walley and Gert De Cooman. Coherence of rules for defining conditional possibility. *International Journal of Approximate Reasoning*, 21(1):63–107, 1999.

[64] L. A. Wasserman. Belief functions and statistical inference. *Can J. Statist.*, 18(3):183–196, 1990.

[65] L.A. Wasserman. Prior envelopes based on belief functions. *Annals Statistics*, 18:454–464, 1990.

[66] S.K.M. Wong, Y.Y. Yao, and P. Lingras. Comparative beliefs. In Ronald R. Yager, Mario Fedrizzi, and Janusz Kacprzyk, editors, *Advances in the Dempster-Shafer Theory of Evidence*. John Wiley & Sons, Inc, Nova York, USA, 1994. ISBN 0-471-55248-8.

[67] Ronald R. Yager, Mario Fedrizzi, and Janusz Kacprzyk, editors. *Advances in the Dempster-Shafer Theory of Evidence*. John Wiley & Sons, Inc, Nova York, USA, 1994. ISBN 0-471-55248-8.

[68] Lotfi A. Zadeh. Fuzzy sets. *Information and Control*, 8:338–353, 1965.

[69] Lotfi A. Zadeh. Fuzzy sets as a basis for a theory of possibility. *Fuzzy Sets and Systems*, 1:3–28, 1978.

[70] Lotfi A. Zadeh. Book review: A mathematical theory of evidence. *AI Magazine*, 5(3):81–83, 1984.

[71] L. Zhang. Representation, independence, and combination of evidence in the Dempster-Shafer theory. In Ronald R. Yager, Mario Fedrizzi, and Janusz Kacprzyk, editors, *Advances in the Dempster-Shafer Theory of Evidence*. John Wiley & Sons, Inc, Nova York, USA, 1994. ISBN 0-471-55248-8.

At each shift of the paradigm, the impossible presents its impeccable
credentials, is overruled and the unthinkable becomes the norm.

— MICHAEL BERG [3]

www.ingramcontent.com/pod-product-compliance
Lightning Source LLC
Chambersburg PA
CBHW060155060326
40690CB00018B/4120